HUNGARY

RUMANIA

CRIMEA

Sevastopol

BLACK SEA

Danube

UGOSLAVIA

BULGARIA

ALBANIA

TURKEY

GREECE

AEGEAN SEA

Athens

Aleppo

SYRIA

CYPRUS

Beirut

PALESTINE

TRANSJORDAN

CRETE

Gazala
Tobruk
Bardia

Mersa Matruh

Port
Said

Gaza

Mechili

Alexandria

Ma'an

Aqaba

Knightsbridge

Beda Fomm

urat

Antelat

Cairo

Suez

Suez Canal

El Agheila
Marada

El Alamein
Alam Halfa

CYRENAICA

E G Y P T

RED SEA

B Y A

C.H. Burttell

Novr. 1978

THE QUEEN'S DRAGOON
GUARDS

THE QUEEN'S
DRAGOON GUARDS

BY
EVERSLEY BELFIELD
WITH AN INTRODUCTION BY
LIEUTENANT-GENERAL SIR BRIAN HORROCKS

LEO COOPER
LONDON

First published in Great Britain, 1978, by
LEO COOPER LTD.
196 Shaftesbury Avenue, London WC2H 8JL

Copyright © 1978 by
The Regimental Association of the Queen's Dragoon Guards

ISBN 0 85052 242 0

Photoset in 10pt Baskerville by Granada Graphics
and printed in Great Britain by
The Hollen Street Press Ltd.
at Slough

To

Brigadier A.W.A. Llewellen Palmer

DSO MC

Contents

	Introduction	xi
	Foreword	xiii
1	The Foundations	1
2	The Continental Wars of the Later Stuarts	7
3	The Long Recessional	13
4	The Revolutionary and Napoleonic Wars	20
5	Service in the United Kingdom	32
6	Colonial Wars	39
7	The Eclipse of the Horse	55
8	Tanks and Armoured Cars – France and North Africa	70
9	Italy	85
10	The Post-War Years	89
11	1st The Queen's Dragoon Guards	97
	Epilogue	103
	Appendices	105

Illustrations

1   *Regimental Uniforms, 1685-1914*   *facing page*   18
2   *The capture of the French standards at Ramiliies*   18
3   *The KDGs at Waterloo*   19
4   *Baggage wagons at Newcastle, 1825*   19
5   *The KDGs in Canada*   34
6   *The Queen's Bays WO's and Sergeants' Mess, India, 1865*   34
7   *Taku Forts, China, 1860*   35
8   *The Capture of King Cetewayo*   35
9   *The Queen's Bays at Montigny-les-Lens*   50
10   *A KDG armoured car in the desert*   51
11   *KDGs escorting Italian PoWs*   51
12   *The Prime Minister visits the Queen's Bays*   66
13   *'General Grant' tanks in the desert*   66
14   *The Colonel-in-Chief presents a new Standard to the Queen's
       Dragoon Guards, 1959*   67

BETWEEN PAGES 90 AND 91

15   *Helicopter and Saladin on exercises in Germany, 1962*
16   *Saracen and Saladin in Aden*
17   *The Regiment on Parade in Germany, 1973*
18 & 19   *In Northern Ireland, 1976*

Illustrations in text

Recruiting poster   *Page* 24
Map: India   42
Battle Honours   96
Maps: Europe and North Africa   *Endpapers*

Introduction

by Lieutenant-General Sir Brian Horrocks

The KDGs and the Bays were both founded by James II in 1685. They are, therefore, two of the oldest Cavalry regiments in the Army. My first encounter with the Regiments was in Aldershot when I was Brigade Major of the 5th Infantry Brigade and they occupied the Cavalry Barracks on the far side of the playing fields. In order to keep the horses fit and to while away the weary hours of the winter we were permitted to hire a Cavalry horse from them for 15/-. They were known as 'fifteen bobbers'. I joined in with the younger officers and we took a disused farm, which our pioneers renovated, in the South Berkshire country, and here on good hunting days as many as 40 young officers would enjoy a day's hunting. We thus got to know each other very well and I was always enormously impressed by their courage and by the care which they took with their horses. On one occasion I can remember the MFH lining us all up and saying, 'If you young officers do not behave yourselves and keep behind hounds I am going home.' It is an interesting fact that this extreme gallantry was one of the causes of their many defeats in the Western Desert. Rommel's Afrika Corps, if faced by the Cavalry, would halt, withdraw behind their formidable anti-tank guns and wait while the Cavalry were mown down. Only then would the Germans advance. Unfortunately, their Mk IIIs were more powerful than our tanks and they also had the deadly 88mm guns. So the odds were loaded against us. This was one of the main causes why we were driven back to the Alamein line.

When I took over 13 Corps in the Desert Rommel was about to launch his main attack on Cairo, and the existing plan was that as soon as he reached the front line our tanks would launch their attack. I determined to avoid this and ordered the Armoured Commanders to dig their tanks into position, covered by anti-tank guns on the flanks, and to have an Armoured Division to block their passage towards Cairo. I remember describing this as 'dog eats rabbit' – i.e. as the Germans penetrated towards Cairo our tanks would issue from both sides and attack their soft vehicles. This was completely successful, though when I explained it to Winston Churchill during one of his visits he was furious and said, 'That is the trouble with you British Generals. Attack! Why not attack? That is the way to win battles!' Actually, the whole thing went according to plan and when the battle was over I asked Montgomery whether I could send a message to the Prime Minister and he said, 'What do you want to say?' I replied, 'Dog ate rabbit!'

The KDGs and the Bays were both involved in this plan and on many subsequent occasions during the Campaign in the Middle East they were under my command. Perhaps the most frightening of all was the Left Hook at Mareth. Montgomery's original plan to attack along the coast had failed owing to the Wadi Zigaou which was flooded and covered by anti-tank guns. The 50th Division were ordered to withdraw. Next morning Monty called a conference and ordered the 1st Cavalry Division and my Corps to do a left hook, join up with the New Zealanders and smash our way through to El Hamma. This we did, but we decided on an unusual attack – to wait till the moon came up and then, with terrific support from the Air Force, to advance down a valley, both sides of which were held by the Germans. I shall never forget this, as it was the most eerie experience of my life. In spite of the moonlight, visibility was nil – there were numerous wadis across the valley which delayed progress – and the Germans took advantage of the situation. However, we got through, reached El Hamma, cut off several German divisions, and Monty described it as the most important Campaign in the Western Desert. Both the Bays and the KDGs played a prominent part in this operation. I always regarded it as a great privilege to have these magnificent Cavalry regiments under my command.

Foreword

In the writing of this brief history I have had help from many sources. It was at the Seigneurie, Dame Sybil Hathaway's house on the Island of Sark, that Brigadier Tony Llewellen Palmer proposed that I should undertake this pleasant task. It was a brave action to entrust the account of two such old and illustrious Regiments and their successor Regiment to one who had never been a cavalryman and knows nothing about horses. Throughout my period of authorship, Tony has been a constant source of encouragement to me and without his generosity the publication of this book would not have been possible. Mr Hollies Smith of the Parker Gallery has always been ready to assist me with his unrivalled knowledge of the history of the KDG in particular and of cavalry matters in general. Brigadier Peter Body, Lieutenant-Colonel Michael Lindsay (who had the unique distinction of twice commanding the KDG) and Major 'Bert' Thorpe provided me with first hand information about the last days of the horse in the cavalry. Major Allsop and Mr Thirkill of the QDG Museum answered my many queries. The late Major-General Beddington read most of the draft of the manuscript and made several valuable comments and suggestions. Finally I would like to thank General Sir Jackie Harman, who inherited the project when he became Colonel of the Regiment, both for giving me so much of his valuable time and for his help.

1

The Foundations 1685-1691

1ST The Queen's Dragoon Guards only began its formal existence on 1 January, 1959, this being the day when the 1st King's Dragoon Guards was amalgamated with the 2nd Dragoon Guards, better known as the Queen's Bays. Thus ended the separate identities of two of the oldest and most distinguished regiments in the British Army. Yet this new regiment emerged like a phoenix; naturally it was fully equipped militarily from the moment of its birth, but it also inherited intact the less tangible, and equally vital, martial qualities which can only be painstakingly acquired through generations of experience in peace and war.

Both the original regiments were founded in June, 1685. They were raised during the brief but stormy reign of James II to help suppress the rising led by the Duke of Monmouth, nephew of the King and illegitimate son of Charles II. Neither of the regiments was ready by 5 July when, at the battle of Sedgemoor in Somerset, this rebellion was crushed. However, the senior, which was then called the Queen's Regiment of Horse, performed its first official duty by providing a mounted escort to bring Monmouth and other prisoners on their journey from Winchester to London. The Third Regiment of Horse, as the Queen's Bays was then named, also started its career by guarding prisoners, and detachments were sent to round up fugitives in the West of England where a bloody penalty was being paid by those involved in Monmouth's rising.

Having crushed the rebellion, James II decided to retain a standing army of 35,000 men. Although this was a small force by comparison with most other European powers, the recent memories of Cromwell still aroused profound suspicion about the retention of any regular army, even of this size. Nevertheless on 1 January, 1686, these two new regiments were established by royal warrant. James had entrusted the raising of the Queen's Regiment of Horse (its full title in 1686 was the Queen Consort's Regiment of Horse) to one of his most senior cavalry officers, Sir John Lanier, and he was to continue as its colonel until killed at the battle of Steinkirk in Flanders in 1692. The contemporary method of recruitment was for men of substance each to raise a separate troop of about 40 to 60 soldiers; these troops were then brought together to be formed into a regiment with their officers. Besides Sir John Lanier, Major William Legge (brother of the Earl of Dartmouth) raised a troop; he was also the lieutenant-colonel of the regiment, thus being responsible for its day-to-day administration, and with Lanier normally being employed as a brigadier-general, Legge would expect to command the regiment in the field. The other seven founding officers who raised troops were Captain Henry Lumley (brother of the first Earl of Scarborough), Lord George Hamilton, Captain Staples, Captain Lewis Billingsly, Captain Charles Nedby (these three coming from other cavalry regiments), Captain George Hastings and Captain Fortrey; all those with troops held the rank of captain.

The smaller Third Regiment of Horse was founded in a somewhat different manner. Initially only four troops were raised, one by Sir Michael Wentworth in the Pontefract area of Yorkshire and the others from the London region by Sir John Talbot, John Lloyd and Lord Aylesbury. The second Earl of Peterborough was appointed Colonel; he was a devoted royalist who had tried to save Charles I in 1648 and had, after the Restoration, been made Governor of Tangier. Almost immediately Peterborough himself raised another troop, as did a Sir John Egerton. Thus initially this regiment had six troops, whilst the Queen's Regiment of Horse had nine.

Of the nine new regiments raised by James II in 1685,

three had been disbanded by 1692. Very confusingly when trying to keep track of them in contemporary accounts, cavalry regiments were then usually called after their colonels, the system of numbering them being introduced later. Although the regiment was the colonel's own property, he normally delegated the actual running of it, as well as the command in battle, to the lieutenant-colonel who, like the other officers, bought their commissions from him.

The men of the Third Regiment of Horse were dressed throughout in crimson, while the facings and linings of the crimson uniforms of the Queen's Regiment of Horse were yellow. Plate no 1 shows the long, heavy jackboots which came up to the thighs. The soldiers' bodies were protected by a heavy leather jacket consisting of a breast-plate and a back-plate fastened together at the sides; this garment was proof against pistol shot and was called a cuirass, hence the word cuirassier for this type of horsed soldier. They wore an iron head-piece, known as a pot. For weapons, each soldier carried a sword, two pistols with barrels 14 inches long and a carbine, or short musket, its barrel being 2 feet 7 inches long. The heavily accoutred 'Horse' or cavalry units rode large strong horses, and were not expected to fight dismounted, unlike the Dragoons who were more lightly clad, carried the longer infantry musket and bayonet and acted as mounted infantry. To anticipate events, the two regiments were only designated Dragoon Guards in 1746 and were therefore conventional cavalry units in every respect until that date.

Returning to the narrative, between 1686 and 1688 at Hounslow Heath, James held several large reviews of his new army, assembling up to 12;000 men, which alarmed many Londoners. He also made journeys through various parts of the kingdom, for which the two regiments provided escorts. But what created most widespread opposition was James's publicly embracing the Roman Catholic religion and some of the measures that he introduced as a result. Among these was the replacement of a number of Protestant officers and men by Catholics. The Queen's Regiment of Horse

3

seems to have been almost unaffected but when the Earl of Peterborough became a Roman Catholic five of the original officers in his regiment left and of their replacements one had an Italian and another an Irish name; it seems likely that some at least of the other ranks were also removed to make way for Catholics.

Late in 1688, William of Orange landed at Torbay in Devon. The two regiments were now faced with their first, and last, major test of loyalty. To help guard him, James kept the regiments near London and they both remained faithful to him. Fortunately for them, James fled the country in December without confronting William in battle and thus the two regiments were spared having to choose sides at this critical juncture in the nation's history. They were therefore able to switch their allegiance smoothly to the new monarch, in whose thirteen years' reign they were to see much active service.

Inevitably the composition of the Third Horse was seriously affected by the change of régime and there occurred a purge of its considerable Catholic, or at least pro-James, element. Two complete troops were drafted in from another regiment that had been temporarily raised by William's supporters and, in addition, 50 more soldiers were brought in from outside. Although impeached, the elderly Peterborough was not executed, but his colonelcy was transferred to the Hon Edward Villiers, eldest son of Viscount Grandison. Presumably many of the officers were replaced, certainly including its Lieutenant-Colonel, John Chitham. The Queen's Regiment of Horse also lost its Lieutenant-Colonel, William Legge, the Hon Henry Lumley taking his place. Otherwise this regiment seems to have undergone few personnel changes.

William's usurpation of the throne created deep resentment in certain circles in Scotland. In April, 1689, the Queen's Regiment of Horse saw active service for the first time, being sent with English and Dutch troops to put down a revolt north of the border. Under Sir John Lanier, one of the commanders of the whole force, this regiment took part

in the siege of Edinburgh Castle, which surrendered in
June. Later in the summer, after the somewhat inconclusive
battle of Killiecrankie, the regiment went into the Highlands to help break up the remnants of the Scottish dissidents.

Meanwhile, William was faced with a far more serious
threat from Ireland where James II had landed in April.
Not only did James obtain widespread support from the
Irish, but he was also assisted by a sizeable French force. In
the autumn of 1689, the two regiments were sent to Ireland
as part of an army which numbered nearly 40,000 soldiers.
They spent the next two years there, suffering their first
casualties while engaged in the dreary round of skirmishes,
ambushes and sieges that characterized this campaign. Both
regiments fought in the battle of the Boyne, when, on 11
July, 1690, William defeated James' Franco-Irish army. A
year later they were both also at the battle of Aughrim in
Galway and sustained heavy losses. The Queen's Regiment
of Horse had twenty-three men killed, while the Third
Horse lost four officers and twenty-six men, as well as having twenty-four wounded; these severe casualties were probably the result of a spirited cavalry charge, possibly the first
made by the regiments, but this had a decisive influence on
the outcome of the battle and the English routed their
opponents. This obscure battle had an important effect on
the war because the French commander St-Ruth was killed
and his death quickly led to the collapse of organized Irish
resistance. The war ended in October with the surrender of
Limerick in whose siege the Third Horse had taken part.
Although a depressing affair, the Irish war must have given
an invaluable apprenticeship to the soldiers of the two regiments, who were certainly inexperienced and probably only
partially trained. They found themselves serving under such
famous generals as Marshal Schomberg (killed at the Boyne
aged 74) whose fighting career, both with the Dutch and
French armies, spanned more than half a century; they also
served under General Ginkel, later created Earl of Athlone.
Furthermore, these newly-raised cavalry regiments were

5

fighting beside many Dutch, Danish and German regiments whose soldiers were hardened campaigners and the Englishmen must have learnt much about the ways of war from the foreigners in this composite army.

2

The Continental Wars
of the Later Stuarts

THE war in Ireland was only a sideshow in the much wider conflict that was being waged to try to prevent the French King, Louis XIV, from dominating Europe. In 1689 a Grand Alliance had been formed between the English, Dutch, Spanish and Austrians and this loose alliance fought the French until the Peace of Ryswick in 1697. In 1691, the French gained some major successes in the Spanish Netherlands, now Belgium, which threatened the whole structure of Holland's security. Hence William desired to end the Irish war as quickly as possible and transfer most of the troops from there to the Continent where Marlborough was already leading 8,000 English soldiers. In August, 1692, the Queen's Regiment of Horse landed in Belgium where they spent the next five years campaigning, mainly round Ghent, Brussels and Namur. Here they were to do battle against some of the finest soldiers in Europe and they may be said to have entered the top league in contemporary warfare. The regiment arrived just too late to be at the Steinkirk where Sir John Lanier was killed. Their lieutenant-colonel, Henry Lumley, who brought the regiment over from England, was promoted colonel, and he commanded the regiment for a quarter of a century. He proved to be one of the finest cavalry officers of his age and it was one in which there were many competitors. Lumley was also the most distinguished soldier this regiment has produced in its long history. At Sawbridgeworth parish church in Hertfordshire a marble memorial was erected in his honour; it says that Lumley 'was in every battle and at every siege, as Colonel, Leiutenant-

Colonel, or General of the Horse, with King William or the Duke of Marlborough, in twenty campaigns in Ireland, Flanders and Germany, where he was honoured, esteemed, and beloved by our armies, by our allies, and by our enemies, for his singular politeness and humanity, as well as for his military virtues and capacity.'

In 1692 the Queen's Regiment of Horse was involved in no action, going into their winter quarters soon after disembarking. As usual in this part of the Spanish Netherlands, siege warfare dominated the campaign, but on 28 July, 1693, one major battle was fought at Neerwinden (also known as the battle of Landen) which is near Liege. William had constructed a strong semi-circular entrenched line about four miles long with Neerwinden village at its centre. Here he stationed 50,000 troops to try to prevent the French advancing westwards. Opposing William was the Duke of Luxembourg who rightly judged that this line was too extensive for the number of troops holding it. Luxembourg had an army of 80,000 and decided to attack, but it took most of a day and some very fierce fighting before the French soldiers managed to break through. When the Allied collapse did occur it was sudden and complete; William's reserves were positioned in Neerwinden itself and consisted largely of English troops, including the Queen's Regiment of Horse. To stem the French onslaught, he personally led a counter-charge and in the affray was cut off and nearly captured, but, after regrouping, a second charge was made in which the Queen's Regiment of Horse was largely instrumental both in saving William and checking the French sufficiently to enable the Allied troops to retreat without disaster. The rest of this campaign was less dramatic, with reconnaissances, endless marches and minor engagements which were mainly connected either with trying to raise a siege or with safeguarding those investing one of the many fortresses in this strategically vital area. Lumley's abilities were already recognized and when the Regiment returned to England in the autumn of 1697 he was a brigadier-general.

The Third Horse were kept longer in England. During

1692, they were employed on patrol duty against the highwaymen who were very active round London, especially in the Blackheath and Hounslow districts. The following year they formed part of the forces kept at home to repel a possible French invasion and it was not until 1694 that they crossed over to the Low Countries where they joined the main army. They returned to England in 1698 but were immediately sent to Ireland where they remained until 1703.

Signed in 1697, the Peace of Ryswick proved but a brief interlude in the wars against France. Early in 1702 the War of the Spanish Succession broke out and lasted until the Peace of Utrecht in 1713. The cause of this conflict was the possible unification of both the French and Spanish thrones under a French king. This would have been the most serious threat yet to the balance of power in Europe, because it would have posed a perpetual menace to the security of all the other states but especially to the English and the Dutch. Soon after the Spanish throne had become vacant on the death of the Spanish king, Charles II, in November, 1700, Louis XIV used French troops to seize parts of the Spanish Netherlands for one of his grandsons, Philip. This precipitated the war. To make matters worse, when James II died in France in the autumn of 1701, Louis immediately recognized his son, the Old Pretender, as James III. To add further tension to the scene, by 1702 it was obvious that William III was dying, but Anne, James II's daughter, succeeded peacefully to the throne and, almost equally as important, Marlborough, William's military protégé, proved to be a genius.

During this war most of the English troops, including the Queen's Regiment of Horse, served under Marlborough, but some small contingents were sent to the Spanish Peninsula where an ill-conducted campaign was waged to try to install the Allied claimant, Charles of Austria, on the Spanish throne. The Third Horse was chosen as one of the two cavalry regiments which formed part of an English force of 6,500 men. The Regiment sailed from Ireland and after many delays arrived in Lisbon early in 1704; they found that the Portuguese authorities could only provide them with half the number of horses they needed and conditions were generally so chaotic that they saw little or no active service that year. The English force was soon halved by transfers to Peterborough's expedition that had captured Barcelona and the Third Horse became the only English cavalry regiment to remain with the mixed army of Dutch and Portuguese on the western side of the Spanish Peninsula. During 1705 the generals in charge of the three national contingents each held command for a week at a time and it was therefore remarkable that any campaigning occurred, but three towns were captured. 1706 marked the climax of this strange, little-known campaign in which the allied troops were led by a French Protestant general, Massue de Ruvigny, later Earl of Galway, while the Franco-Spanish army was commanded by an English Catholic, Marlborough's nephew, the Duke of Berwick. With 19,000 men, mainly Portuguese, Galway outnumbered Berwick who fell back, and, in June, 1706, the Allies entered Madrid proclaiming the Austrian Charles as king of Spain. This was an empty gesture as Charles was still only slowly approaching from the opposite direction, and, by the time he had reached Madrid, the ill-disciplined Portuguese troops had so incensed the local population that they changed their allegiance to support Philip, the French claimant. Berwick's army had also been reinforced and he drove Galway eastwards to the coast at Barcelona. Here in Catalonia Galway was assured of the backing of the British navy and the inhabitants were friendly. Thus in six months the Third Horse had marched almost across the Spanish Peninsula. In 1707 they fought at Almanza where in a dashing charge they successfully helped to break up two French battalions, but, with the Portuguese cavalry having left the field, the battle turned against them. A French counter-attack inflicted heavy losses on them and the surviving Allied troops were forced to withdraw to Catalonia again. Here they spent the next two years in desultory fighting. The climax of this see-saw campaign came in 1710. The year opened well with a major victory over the opposing French and Spaniards at Almenara where the Third Horse formed part of Stanhope's cavalry which routed the enemy. A harder fought success was later gained at Saragossa and Philip's force was defeated. In September the Allied army, with Charles at their head, re-entered Madrid. There they waited for Portuguese reinforcements which never arrived. By November their position had become untenable and they were compelled to march back through hostile territory to try to reach Catalonia again. By now Philip's army had been greatly strengthened and they cornered just over 2,000 British troops at Brihuega. After a brief but bitter siege, they surrendered. This was the only time that this Regiment was captured; an exchange was, however, arranged with the French and, after some months, the Third Horse were repatriated in the autumn of 1711.

In the War of the Spanish Succession, the Queen's Regiment of Horse served abroad continuously from 1702 to 1714. They formed part of Marlborough's forces and distinguished themselves in all his major victories and campaigns. (As there are so many excellent accounts easily available, this war will not be described in any detail and only the outstanding episodes will be mentioned.) When, on the long march from the Netherlands to the Danube in June, 1704, Prince Eugene of Savoy met Marlborough for the first time he commented on the fine state of the English cavalry. Eugene's tribute provides a wonderful picture of these soliders. He said of them: 'I never saw better horses, better clothes, nor finer belts and accoutrements; yet all these may be had for money; but there is a spirit in the looks of your men which I never saw before, and which cannot be purchased. It is an earnest of victory.' Two months later, at Blenheim, the Queen's Regiment of Horse was posted on the Allied left flank and thus in the thick of the fierce struggle that took place to gain the village of Blenheim itself. At the end of the day Lumley was one of the commanders who led the cavalry charges that finally disintegrated Tallard's army, driving the gallant French soldiers into the Danube. In 1706 at Ramillies the English cavalry were on the right of the line. They

11

were not employed until the very end of this battle when, under Lumley, they swept into action at the gallop to cut the French line of retreat, the Queen's Regiment of Horse playing a particularly prominent part in this role and during the harassment of the enemy's withdrawal. The pursuit went on till the early hours of the morning of 24 May. The campaigning season lasted from May to October and in each of the years from 1704-7 the Queen's Regiment of Horse may well have covered about 1,000 miles. The winter months were spent quartered in Holland or in Flemish towns such as Ghent; reinforcements and remounts were sent over and the regiment prepared itself for the next round of fighting.

Although no order of battle exists for Oudenarde, it is likely that the Queen's Regiment of Horse were present. On that hot July day in 1708 Marlborough's troops had to make a long approach march before reaching the battlefield where they were immediately sent into action. Lumley's small force of English cavalry were given a strange and frightening role when they arrived late that afteroon. They were posted to protect the right wing where a very large body of French troops seemed poised to descend on the Allied foot soldiers who were locked in fierce and close combat. For several hours Lumley had to sit motionless at the head of this small force waiting to repel a massive French attack which fortunately never materialized. The day after this great Allied victory, Lumley led an unsuccessful pursuit, but the French escaped in good order to the fortress of Ghent. Later in 1708, the Regiment helped cover the allied armies during the bloody siege which eventually led to the capture of Vauban's greatest fortress, Lille. In 1709, the Regiment fought at Malplaquet where they vigorously pursued the French troops who were fleeing from the battle. In the concluding years of this long campaign they were employed in routine duties.

3

The Long Recessional
1715-1783

BOTH Regiments spent most of these years in the United Kingdom, with regular tours in Scotland and Ireland, but they were fortunately not involved in suppressing the Jacobite Risings in 1715 and 1745. They changed their names several times. The new king, George I, being a widower, in 1714 the Queen's Regiment of Horse was renamed the King's Own Regiment of Horse. The Third Regiment of Horse became the Princess of Wales' Own Royal Regiment of Horse, but on the accession of George II, in 1727, their name was again changed, this time to the Queen's Own Royal Regiment of Horse. Finally in 1746, both Regiments were turned into Dragoons after which they retained their titles until amalgamation. The one was established as the First or King's Regiment of Dragoon Guards. The other was called the Second or Queen's Dragoon Guards, but, after 1767, when it was officially mounted on bay horses, it became popularly known as the Queen's Bays. (This name was not officially recognized until 1872.) The reason for the change in status from Horse to Dragoons was to economize, the rates of pay for Dragoons being lower, and over £7,000 per annum was saved with the KDGs and rather less with the Queen's Bays which was smaller, having six troops to the KDG's nine. The title of Guards was granted to all the regiments which were transferred from Horse to Dragoons and they were also given seniority over all the existing regiments of Dragoons. Those soldiers wishing to be discharged were given a small gratuity, while those agreeing to serve on received a bounty of £3.

The total sum of money allotted for the running of the regiments (less legitimate and illegitimate deductions made by the Paymaster-General's department) was paid directly to the Colonel who would normally expect to make some profit from his purchase. This system was wide open to abuse and there can have been few regiments where an unscrupulous colonel was not appointed at some time or another. The position was made worse by rates of pay remaining unchanged almost throughout the eighteenth century. With the increasing price of forage the other ranks and junior officers found it almost impossible to make ends meet. Admittedly some recognition of the rising cost of equipping a regiment was recognized in 1767 when the purchase price allowed for a troop horse was raised from 20 to 22 guineas. Recruits became more and more difficult to obtain, while the policy of offering a bounty to a recruit on joining led to widespread desertions and a class of men known as 'bounty jumpers'. As no barracks existed outside London, one must visualize the regiments continually on the move and fragmented on a troop basis which made for easier billetting, in an inn in the winter or camping out in the summer. In September, 1737, Capt Philip Browne of the KDGs wrote from Market Harborough, 'I relieve the Officer the 1st of October . . . and march the troop to Lutterworth . . . there being a horse fair to be kept there . . . and holds a fortnight, and we go for the convenience of the inhabitants . . . and then if nothing unforeseen happens go to Northampton for the winter'. A little later he spoke of being at Kettering. In 1740 he wrote from 'the camp near Newberry, very High Winds and Hard rains for some time past. The tents are now never dry, the Men are continually wet, and the Horses stand footlock deep in Water frequently . . . The only way that is possible to endeavour to prevent catching Cold is to drink more Wine than is necessary at other times.'

The Queen's Bays were also continually on the move. There being no proper police force, troops of cavalry were employed as a mobile law-enforcing arm. Much of the Regiment's time was spent in the coastal areas where they assisted the hated excise men in trying to curb smuggling. With the growing tensions built up by the Industrial Revolution, they were often called upon to help the local magistrates quell disturbances. In 1769 the Queen's Bays were stationed in the Lancashire coalfields to suppress disorders there; in 1791 the KDGs put down riots in Birmingham. Although it made them unpopular, it would seem as if the cavalry became skilled in restoring order, probably using the flats of their long swords with considerable effect. They seem to have done their duty without ever resorting to the sort of indiscriminate violence that characterized the efforts of the untrained Yeomanry at Peterloo in 1819.

These duties must have left little opportunity for more serious military training, but the Queen's Bays was one of the few regiments that established a riding school for itself. Operational efficiency was further reduced because directly after every campaign the number of men in each regiment was cut by about one third. In 1779 the light troops in each regiment, which had been added in 1755 and trained for skirmishing and scouting duties, were transferred to form part of the 19th Hussars. Considering all these handicaps, the Regiments acquitted themselves creditably during the two continental wars in which Britain was engaged between 1714 and 1793.

The first of these was the War of the Austrian Succession. In 1743 the King's Own Regiment of Horse (KDGs) was hurriedly brought up to strength and its 535 officers and men sent across the Channel. Writing from Ghent on 12 September, Captain Browne told of arriving at Ostend, after

a very disagreeable and fatiguing Passage of twelve days. That morning we disembarked the Horses upon the Sands, the tide being half way up their legs, where we remained till Eight at Night, and then marched to Bruges (11 miles) where we arrived at One a Clock in the Morning. Neither Officers, Men nor Horses, having had any refreshment that whole day, we was obliged to threaten to brake open the doors of my Inn,

15

before we could get admittance, after which enquiring for a bed that was said we must lay in the Kitchen, but we took the Liberty to take possession of the Land-lords[1].

In 1743 the Regiment fought at Dettingen under General Honywood and two days later Captain Browne wrote a somewhat breathless account of this battle:

For several hours we stood the cannonading of the enemy, from several batteries they had erected, which commanded the line of march, so as not only to annoy us, but frequently went beyond us . . . and so soon as men or Horses was killed they closed again, & at the same time we could see that as our cannon played upon them, they sett up a gallop in great disorder. They began by cannonading us upon our march; there first fire was at eight a clock in the morning; they had been marching all night, & had passed the Maine, and drawn their troops up on this side, in order to prevent our joyning six thousand Hessians, & eight thousand Hanoverians, which came to this ground last night, they took possession of a village where they crossed over, which was a pass we must fight through or retire back again – immediately we formed & marched in line of battle . . . about a mile before we came to the ground, where we engaged, our eyes was presented with numbers of dead bodies, & some that was shot and slain & not expired, which we could not help riding over & passing through; I saw numbers that the foot put an end to by firing their pieces in their ears. Before the left of the brigade of Horse was formed, the Gens-Arms, [presumably the Gendarmes, a French Household Cavalry Regiment] the best troops of France, advanced to attack us, & a battery of their cannon flanked us; upon their advancing to attack Genl

[1]*Journal of the Society for Army Historical Research,* vol v, Jan-March, 1926.

Honywood's & General Ligonier's regiments we marched forward and met them sword in hand; at the same time their cannon ceased, & they flanked us on the left with their foot; then we engaged & not only received but returned their fire: the balls flew about like hail, & then we cut into their ranks & they into ours ... Cornet (George) Allcroft, who was near me, was killed, and the standard which he bore was hacked, but we saved it ... Our Squadron suffered most, we being upon the left.

I did not receive the least hurt but my left hand & shirt sleeve was covered with blood, which must fly from the wounded upon me. Providence greatly favoured me that as their was an engagement we was in the thickest of it & was my kind protector, had not the English foot come to our relieve we had been all cut to peices, the Gens. Arms being nine deep & we but three, after which we rallyed again & marched up to attack them again, but before we was ordered the French had retired & the English, Hanoverians & Austrians remained masters of the field ...We had nothing to eat nor drink, & we quenched our thirst by the rain that fell upon our hats, & we had nothing at all for our horses.

In 1745, at Fontenoy, the British cavalry charge was performed with gallantry, but this ended in failure. In these battles and other minor encounters, the Regiment lost three officers and fifteen men killed and over twice that figure wounded.

The Seven Years War lasted from 1756 to 1763. In 1758 a series of amphibious landings took place on the French coast. Consisting of nine infantry battalions and the Light troops from nine cavalry regiments, including that of the First Dragoon Guards, this force was commanded by Colonel Eliott, later commander of the Gibraltar garrison during the Great Siege of 1779-83. Their first operation was the occupation of St Malo where shipping and stores were

17

burnt. In August the same expeditionary force made a more daring raid in which they seized the naval base of Cherbourg, destroying most of the fortifications there, as well as blowing up the shipping in the harbour before being successfully re-embarked; they thus anticipated the Normandy landings by nearly two centuries. A further raid, this time on the Breton coast, met with much stiffer resistance and the expedition was hard put to return to its vessels.

In the main theatre of war the King's Dragoon Guards were present at the battle of Minden in 1759 when the cavalry, under Lord George Sackville, were prevented from joining in the fighting, despite orders to the contrary from the Commander-in-Chief. The following year the Bays joined the KDGs and both fought at Corbach which was memorable for a desperate charge by the KDGs in which they and another regiment helped save the army which was in full retreat from the French; the Regiment lost forty-seven killed. Soon after this the two Regiments served for the first time in the same brigade at the battle of Warburg. Although a minor victory, Warburg was important because the British cavalry, under the dashing Marquis of Granby, restored their reputation which had been tarnished by Sackville's behaviour two years previously. The main feature of this battle was the two-hour approach march by which Granby surprised the French. Only pausing to draw up the brigades into two lines, the Dragoon Guards being in the front line with the KDGs on the extreme right, Granby led the charge at a gallop, overthrowing the enemy who fled across a nearby river with the British still in pursuit. The war dragged on for another three years and the regiments took part in many of the encounters of which Groesbenstein in 1764 was the most significant. Here Granby again surprised the French, but this time they retreated rather than risk a battle. The prize, as at Warburg, was the great fortress of Cassell and, after some more skirmishing in which the cavalry played a major role, it surrendered at the end of the year.

Early in 1767 both regiments returned to England. The Bays then mustered 15 officers, 325 men, 31 officers' ser-

1 Regimental Uniforms, 1685-1914

2 The Capture of the French Standards and Kettledrums at
Ramillies, 1706. From the painting by R. Hillingford.

3 The King's Dragoon Guards passing the Duke of Wellington before Waterloo, 1815.

4 Baggage wagons of the 1st Dragoon Guards outside St Nicholas Church, Newcastle, 1825. From the painting by Henry Parker.

vants and a mysterious band of 31 women, but by 1783 their
strength had dwindled to a mere 90 men and 90 horses.

4

The Revolutionary and Napoleonic Wars

AFTER 1792 a programme of barrack-building was begun which largely ended the inefficent and unpopular system of billeting small scattered groups of troops in ale-houses. In 1796 to the establishment of each cavalry regiment was added a veterinary officer who was paid £128 p.a. Alterations were made to the uniform and in 1798 the muskets and large pistols were withdrawn and the men were issued with a carbine and a smaller pistol. The length of hair on men and horses also came in for severe pruning; for over thirty years the horses were long-tailed, but after 1799 the horses' tails had to be docked, and after 1809 the men were no longer permitted to wear their hair in a queue falling ten inches below their collar!

During the latter part of the eighteenth century three very able and original officers joined the Regiments. The first of these was Banastre Tarleton who was commissioned into the KDGs in 1775 and immediately volunteered to fight in the American War of Independence. From his arrival there in 1776 until his capture at the surrender of Yorktown in 1781, he proved to be a most able and energetic commander, frequently surprising and defeating far larger American forces. *The Dictionary of National Biography* states that he 'was a born cavalry leader, with great dash, as such he was unequalled in his time'. He was later promoted Lieutenant-Colonel, but soon gave up his military career to become a Member of Parliament. The most famous of the three was the Guernseyman, John Le Marchant, who joined the Bays as a Lieutenant in 1789 and stayed with them for four years.

Subsequently he founded the Royal Military Academy, Sandhurst, becoming its first Superintendent. He was killed at Salamanca in command of the Heavy Cavalry Bridage whose magnificent charge was largely responsible for Wellington's victory over the French. It was said of Le Marchant that he was 'one of the few scientific soldiers in the cavalry arm'; he was also a gifted water-colourist. Least well known of the three was Robert Long who was commissioned into the KDGs in 1791, staying with the Regiment for five years. In those days when promotions had usually to be purchased, many of the more ambitious officers moved from regiment to regiment as and when vacancies occurred. Like his friend Le Marchant, Long obtained a Lieutenant-Colonelcy cheaply, for £2,000, but this was in one of the emigré regiments, Hompesch's Mounted Rifles. Recognising his outstanding ability, Sir William Pitt, then Colonel of the KDGs, offered Long, in 1799, the Lieutenant-Colonelcy of the Regiment. Then only twenty-eight years old, Long refused it, on the grounds that he preferred not to command officers who had only recently been senior to him. It might be claimed that he was the first QDG officer, since he commanded the Bays from 1803 to 1805. He was Adjutant-General for the Walcheren Expedition before taking part in Sir John Moore's famous march to Corunna where he was at Moore's deathbed. From 1811 to 1813 he was a major-general with Wellington in the Peninsula. He was at Salamanca, where he wrote of Le Marchant's death, 'He was one of the few faithful and esteemed military friends left me,' continuing, 'What will become of his ten orphans, I known not. They have only God and a few friends left for their protection.' Long returned to England in 1813, and although never again on the active list, was promoted lieutenant-general in 1821.

For most of the years between 1792 and 1815 Europe was engulfed first by the French Revolutionary Wars and then by the Napoleonic Wars. During this prolonged conflict, Britain heavily subsidised a series of alliances against the French. Partly as a result of this and partly from geographical difficulties, her land forces were only intermittently involved in the fighting on the Continent.

The KDGs and the Bays took part in the opening stages of the war. Having been rapidly reinforced, both Regiments were sent out to the Low Countries with the Duke of York's expeditionary force and displayed their accustomed dash and valour. This ill-managed campaign lasted from 1793 to 1795 and consisted of some sieges and several minor battles; the most important of these being at Tournai, (also known as Beaumont), where, in May, 1794, the Bays formed part of a British cavalry group whose successive charges so demoralized the French that their troops retreated with considerable loss. A little earlier the KDGs had shown equal gallantry in turning the flank of a French army at Le Cateau. Long had an unfortunate experience in this encounter. Nearly all the leading squadron of the KDGs were unseated when they plunged into a very deep ditch. His horse was wounded, in his own words, 'by musket shot which a scoundrel, whom I pardoned the instant before on condition of his laying down his arms, fired at me. My sword soon gave him the reward which such Criminal Ingratitude justly merited.'

Soon afterwards the alliance began to disintegrate and for the rest of the year the British conducted a slow and expensive retreat across the Rhine. During most of 1795 they remained in Germany without fighting, being transported back to England from Bremen at the end of that year. It is unlikely that Long would have got out with his elaborate camping equipment which had cost him £47.17.4; besides a large quantity of bed linen, a stool and a table, it included a marquee and a tent as well as a round tent for his servant which in an emergency he reckoned he could use himself. The total casualties in this miserable campaign were heavy, the six troops of the KDGs alone having seventy-six men killed and losing 353 horses.

For almost the whole of the Napoleonic War period both the Regiments were kept at home for defence duties, including the inevitable spells in Ireland. In 1809 a detachment of the Bays was selected for the ill-fated Walcheren Expedition, a part of the world most unsuitable for cavalry operations, and although Flushing was eventually captured, the terrible toll of disease soon led to the recall of this force.

Leading a recruiting party was a task which most junior officers had to perform at some time and in 1792-3 Lieutenant Long recorded some of the regimental orders given to him. The recruits were to be confined to Protestants who were British born; excluded from this category were apprentices, seamen, marines, militiamen, collliers, stragglers or vagrants as well as possible deserters. The medical requirements were stringent; all were to be rejected who suffered from 'fits, rupture, broken bones, sore legs, scaled head, blear eyes or running sores and were not perfectly straight, well-featured, in every way, well-made and not heavy limbed.' The age range was 17 to 30, but 'any fine boy' over 16 might be considered. Long was given £5 per recruit, but £2.18s. was to go directly to the man. A special bounty of 14 guineas was to be paid out for any recruit who was over 20 and at least 5 feet 7 inches tall. Set such high standards, it is not surprising that, in two months, only eight men were obtained at a cost of £144 expenses, excluding the pay of Long, the sergeant and the six troopers who formed the party!

Even making all allowances for a wartime emergency, the glowing picture of military life promised by the poster overleaf (circa 1800) seems to contravene the most rudimentary standards of advertising honesty.

After Napoleon's return from Elba in 1815, the KDG strength was increased to 1148 officers and men, its greatest ever. At the end of April 523 of the officers and men, under Lieut-Colonel Fuller, arrived at Ostend to join Wellington's army.

On 15 May Charles Stanley, 'Privert Kings Dragoon Guards' wrote a remarkable letter to his cousin from 'Brusels Flemish Flander's. Although some privates could probably have been able to read a little, very few could have managed to write home, let alone in such a good hand. Fortunately Stanley expressed his thoughts as he would have spoken

FIRST, or
KING's
Dragoon Guards
Commanded by
Gen. Sir. Wm. Augustus Pitt, K.B.

A Few
DASHING LADS
ARE NOW WANTED
To complete the above well-known
Regiment to a New Establishment

Any YOUNG MAN who is desirous to make a Figure in Life, and wishes to quit a dull laborious Retirement in the Country, has now an Opportunity of entering at once into that glorious State of Ease and Independence, which he cannot fail to enjoy in the

KING'S DRAGOON GUARDS

The Superior Comforts and Advantages of a Dragoon in the Regiment, need only to be made known to be generally covetted.

All Young Men who have their own Interest at Heart, and are fortunate enough to make this distinguished Regiment their Choice are requested to apply immediately to

Serj. TIBBLES, at the Angel Inn, Honiton,
where they will receive
THE HIGHEST BOUNTY
And all the Advantages of a Dragoon

As Recruits are now flocking in from all Quarters, no Time is to be lost; and it is hoped that no young Man will so far neglect his own Interest as not embrace the glorious Opportunity without Delay

N.B. This Regiment is supposed to be mounted on the most beautiful, fine, active black Geldings this Country ever produced

The Bringer of a good Recruit will receive *a Reward of THREE GUINEAS.*

them, though his spelling now needs some effort to get
accustomed to, and there is a complete absence of punctua-
tion.

Dear Cufson I take this Oppetunety of Riting to you
hoping this will find you all In gud helth as it leves me
at Prefsent I thank God for It I have ad a Very Ruf
march Since I sow you at Booton (where his cousin
lived) we am onley 15 miles From Mr Boney Part Har-
mey wish we Expect To have A Rap at him Exerry Day
We have the Most Cavilrey of the English that Ever was
None at one time and in Gud Condishon and Gud
sperrits we have lost a few horses by hour Marshing I
have the Plesure to say my horse Is Better Everry Day
wish i think im to be the Best frend i have at Prefsant
there is no dout Of us Beting the Confounded Rascald
it ma Cost Me my Life and a meaney more that will
onley Be the forting of War my Life i set no store By at
all this is the finest Cuntrey Exer is So far before England the Peepel is so Sivel ...
Langwigs we do a grate del my makin moshins – We
have one gud thing Cheap that is Tobaco and Everry-
thing a Cordnley Tabaco is 4d Per 1b Gin is 1s 8d Per
Galland that is 2½ Per Quart and Everything in Perpo-
sion hour alounse Per Day is One Pound of Beef and
Pound and half of Bred half a Pint o Gin But the worst
of all we dont get it Regeler and If we dont get it the
Day it is due we luse it wish It is ofton the Case I assure
you ... I hope you never will think of being a Soldier I
Asure you it is a Verry Ruf Consern I have Rote to my
Sister Ann and I ham afraid she thinks the trubel to
mush to anser ...
I have not ad the pleasure of Ling in a Bed since In the
Cuntrey thank God the Weather is fine Wish is in hour
faver we Get no Pay at all onley hour Bed and mete
and Gin we have had 10d Per Day soped from us wish
we shal Reseive wen six months is Expiered I think
God i have a frend with me ...

25

I hope you will Excuse my Bad Inditing and Spelling[1]

In a letter dated 3 July, 1815, Troop Sergeant-Major Page of the KDGs described events just before the battle;

We were very comfortably situated in Flanders, in good quarters wanting for nothing, till the morning of Jun 16 when at daylight we received a sudden order to march . . . We marched this day about 40 English miles and slept in the open corn-fields, our horses saddled ready to mount at a minute's notice, the French being in a wood close by us. On the morning of the 17th at daybreak firing again commenced. So far it was what we call skirmishing – however, in the course of the morning the French marched out of the wood in numbers double to our own, heavy firing of cannon and musketry commenced and at the same time there was one of the heaviest storms of rain ever known accompanied with thunder and lightening. . . . The fall of rain was so very heavy on the 17th that in the fields, which were covered with corn, our horses sunk in every step up to near the hock, therefore our cavalry could do but little this day. Firing on each day ended when darkness commenced and we remained in the open fields, our horses saddled and bridled the whole night for fear of an attack before morning.

It is out of my power herein to express our situation – our boots were filled with water, and as our arms hung down by our side the water ran off a stream at our finger ends. We remained in this situation the whole of the night halfway up to our knees in mud. Firing commenced the next morning, viz. the 18th, at daybreak which made the third day. What seemed worst of all during these three days, we could draw no rations, consequently we were three days without anything to eat or drink.

[1]This letter is in the National Army Museum.

Sergeant-Major Page also gave a very moving account of his experiences at Waterloo:

> After the action commenced we began to get dry, and as the rain ceased we wrung out our clothes, put them on again, and very few of them have been pulled off since . . .During the morning part of the day, the whole of the British Cavalry were in columns behind the Infantry and Artillery. We lost many men and horses by the cannon of the enemy. While covering the infantry we were sometimes dismounted in order to rest our horses and also when we were in low ground so that the shot from the French might fly over our heads. Whilst in this situation I stood leaning with my arm over my mare's neck when a large shot struck a horse by the side of mine, killed him on the spot and knocked me and my mare nearly down, but it did us no injury.
>
> Soon after this our Brigade was mounted, which Brigade is composed of four Regiments – 1st Life Guards, 2nd Life Guards and Blues four troops each, and our Regiment eight troops – our Brigade is commanded by Lord Edward Somerset. At this time the French seemed determined to get possession of a piece of ground where part of our line was drawn up, accordingly they brought forward very heavy columns of infantry and strong bodies of heavy cavalry, and our Brigade was ordered to form line immediately. Now comes the most bloody scene ever known – the French infantry and cavalry came boldly into the bottom of a very large field while we were formed at the other end, they charged our infantry and as soon as they showed themselves to our front the word charge was given for our Brigade by Col Fuller, who soon fell at our head – deeply regretted. However, we overturned everything, both infantry and cavalry, that came in our way, such cutting and hacking never was before seen. When the French lines broke and turned and ran, our Regiment being too eager, followed the French Cavalry

while the cannon and musketry was sweeping our flank. Many fell and our ranks suffered severely – the Duke of Wellington, with tears, it is said, when he saw us so far advanced among the French, himself said he never saw such a charge, but he was afraid very few of us would return, – his words were too true . . .However, of the 7,000 Frenchmen wearing armour very few left the field. They were very fine men but they could not look us in the face, and dreadful was the havoc we made among them. We lost but few men by their swords, it was the grapeshot and the musketry that cut us down before we got amongst them. We had to charge to meet them so far over heavy ground that many of our horses were stuck in deep mud. The men were obliged to jump off, leave them and seek their safety away from the cannon fire.

My mare carried me in famous style, she got a light wound in her off hind leg by a French Lancer. I was after a French officer who was riding away from me, I came up to him and he thrust his lance at me, I turned it with my sword, it glanced down and cut my mare below the hock of the off hind leg. I was struck by a musket shot on the left thigh, but it was prevented from doing me harm in a singular manner, which was as follows. They day before my sabretasche, which is a kind of pocket made of leather, had one of the carriages broken and in order to keep it safe it was taken up very short and lodged on my left thigh. The pocket being very full of books and other things prevented the shot from going right through when it struck me. This shot would have fractured my thigh-bone had not he sabretasche prevented it.[1]

On 9 July, Sergeant Stubbings, of the KDGs wrote more briefly to his father:

[1]This letter is in the possession of the Regiment.

I take the first opportunity that lays in my Power of informing you that I ham in Good Helth after the very sharp Ingagement which tooke place on the 18 June and Dear father it is a wonder that I escaped without reciving any injury for Whee was verry much Exposed to Danger both on the 17th and 18th of June ... Whee expect all the fitting is over which I hope it is for it is Dredfule to relate the scens I saw on the 18th The field for Miles around was covered with wounded and Slain and in some Places My Horse Could not Psfs (sic) without Trampling on them I am Sorry to inform you that Charles Stanley (the writer of the first letter) fell on that Ever Memorable Day the 18th June fiting Manfully in the Defence of his Country.

The main action described in this letter was the charge of the two heavy cavalry brigades. The KDGs was much the largest regiment in Lord Edward Somerset's 1st Heavy Cavalry Brigade; (it is now often erroneously referred to as the Household Cavalry Brigade and thus the KDGs' part in this charge goes unrecorded in many accounts of the battle).[1] The KDGs were in the centre of this Brigade which was positioned on the right of the main Brussels-Charleroi road; Ponsonby's 2nd Brigade was on the other side; the whole force was commanded by the Earl of Uxbridge. At about 2.20 pm Somerset's brigade was unsuccessfully attacked by the cuirassiers of Dubois' Brigade, on the left-hand side of the road D'Erlon's Infantry Corps had advanced against Picton's Corps and had also been repulsed after some ferocious fighting. Seeing that an opportune moment had arrived, Uxbridge ordered both cavalry brigades to charge and finish off the French attack. The cavalry did their work only too well and despite Uxbridge's

---

[1].Fortescue mentions the KDGs as being involved in the famous charge, but in his order of battle omits the KDGs in the list of regiments forming the 1st Cavalry Brigade. Hon. J.W. Fortescue, 'History of the British Army', vol X, p 428.

efforts to recall them the cavalry became carried away by their success against the fleeing French troops, and continued their charge down the slope and up the far side where they ran into strong French positions. Although they silenced thirty guns, this headlong assault soon petered out against the greatly superior enemy forces and the British cavalry suffered very heavily, only a few scattered groups returning to their own lines. During the remainder of the battle, the depleted heavy cavalry – both brigades were now inextricably mixed up – fought several more engagements assisting the infantry to drive off the fierce French attacks. At the crucial juncture in the evening, before the Prussians had arrived and when the Allied centre looked like crumbling, the Heavy Cavalry were strung out in a single line to deceive the French into thinking that their total strength was more than about two squadrons. Finally, as the French broke, the remnants of the two cavalry brigades once again charged the retreating French soldiers.

The 'butcher's bill' had been terrible. Of about 520 men of the KDGs who had taken part in the battle, 129 had been killed, including their Commanding Officer, and 134 wounded; in addition 269 of their horses were killed. In his despatch, Wellington wrote, 'Lord Edward Somerset's brigade . . .highly distinguished themselves.' In recognition of its distinguished services, the Regiment was given the right to inscribe the word 'Waterloo' on its Standard and accoutrements. All those who fought at Waterloo were granted an extra two years of service towards their pay and pensions and received a silver medal, the first occasion one had been issued to everyone who was engaged in a battle.

The KDGs entered Paris on 7 July, thus being among the first British troops ever to have done so as conquerors. On 24 July they and the Bays, who had arrived just too late to fight at Waterloo, formed part of the Allied Army reviewed there by the Duke of Wellington. Also present were the Czar, the Emperor, the King of Prussia and Louis XVIII of France.

Writing from near Paris in January, 1816, Sergeant-Major Page described the typical disappointments of the military.

> My wife has not joined me yet, nor cannot do till I can get her a passport. She came as far as Dover, but was obliged to return, no more soldiers' wives being allowed to come over; the Duke of Wellington will allow only six to every hundred men. I am so fond of being in the Army in some respects that I should be sorry to leave it; my situation in it is a respectable one, but you see what troubles I am exposed to. No-one knows the soldier's troubles, fatigues and dangers but themselves. In regard to my wife and family, I could support them in a very comfortable manner, but cannot get them to me, but have to support them to so much disadvantage where the necessaries of life are so very dear to what they are in France.

The KDGs returned home in May, 1816, but the Bays stayed on with the Army of Occupation in France till 1818.

5

Service in the United Kingdom

FROM 1818 to 1838 both Regiments reverted to their traditional role of maintaining law and order on each side of the Irish Sea. In 1821 the Bays were reduced to a peace-time establishment of six troops (three squadrons), each of forty-eight privates. (The term 'trooper' is of late nineteenth century origin.) The full regimental strength was 363 all ranks and 253 troop horses, the total number of horses would be about 300, counting all the officers' and headquarters staffs' animals. This being a stormy two decades, the Regiments were continually on the move. For instance the KDGs returned from Ireland in 1822 and, after landing at Liverpool, were dispersed into quarters in Manchester, Sheffield and Nottingham; in 1823 they went to Scotland, being quartered in Edinburgh and Perth. 1824 saw them in England again with detachments in Carlisle, Newcastle upon Tyne and Leeds and in the following year they marched to a large cavalry review by the Duke of York at Hounslow in which the Bays also took part, and moved on to Canterbury, Deal and Shorncliffe. Early in 1826 two troops were dispatched to Norwich, while the rest of the regiment was sent to Leeds, Burnley and Blackburn, a region where civil disturbances were rife. In 1827 the KDGs went back to Scotland, being quartered in Edinburgh, Glasgow and Perth; in 1828 they were ordered south of the Border to York, Carlisle, Beverley and Newcastle upon Tyne, and at the end of this year were centred on Manchester, but were ceaselessly called out to assist the civil authorities in Macclesfield and Rochdale, as well as in Manchester itself. In 1829 they were ordered to

Ireland which was always simmering with trouble and where the Bays had spent most of this period. In 1835, the KDGs were in Wolverhampton and had to disperse rioters during the elections there.

By about 1840 the character and the life style of the British cavalry regiments had altered and they entered a new phase which, with some modifications, was to continue until the twentieth century. Changes occurred gradually and extraneously and these tended to alter the duties of the cavalry to their advantage by making most of them more professionally competent.

Four main factors contributed to the cavalry regiment becoming a more homogeneous and unified organization. First, the formation of permanent police forces helped put an end to cavalry detachments being misemployed as peripatetic law-enforcing bodies. Secondly, the industrial unrest and revolutionary fervour was now abating, thus removing fears of wide-spread violence and this, in turn, reduced the likelihood of military intervention in civil affairs. (These two factors applied far less to Ireland where a significant proportion of the cavalry was always stationed.) The third factor was that sufficient numbers of barracks had now been built to allow regiments to be kept together, although often in very cramped conditions; this concentration of its personnel also assisted training. The final and perhaps most important cause was the expansion of the Empire which, after the Mutiny, included India. The continuing problems of securing peace there and elsewhere, as well as the protection of Britain's global interests, resulted in nearly all regiments having regular overseas tours, thus fostering a close-knit community with a high degree of regimental spirit.

Published in Montreal in 1840, the Standing Orders of the KDGs give an excellent picture of the way in which the life of a cavalry regiment was organized. Consisting of nearly seventy pages, it was issued to every officer and, in the words of the Commanding Officer, Lieut-Colonel the Hon George Cathcart, 'was framed in accordance with Her Majesty's General Regulations.' The booklet is divided into four chapters, the first two deal, at some length, with the General Duties of the Officers, NCOs and Men, the third covers the Ordinary Routine of Regimental Duties and the fourth, something of a rag-bag, is called Interior Economy.

Much was expected of the Troop Commander:

> It is absolutely necessary that every Officer command-ing a Troop should be sufficiently acquainted with it, to be able to answer, at any moment, every question that may be asked respecting the men and horses belonging to it, without referring to his Troop Sergeant Major . . . There is nothing that requires the attention of Captains more than the care of their recruits, who ought to be treated with every encouragement, compatible with duty, by the Non-Commissioned Officers and Men of the Troop, and should be each attached to a smart sol-dier . . . one who is likely to teach him honest and soldier-like habits and not lead him into drunkeness or dissipation.

Concerning officers' language:

> It is strictly enjoined that all orders should be peremp-tory, but on no account accompanied by an oath or invective.

It was recognized that Officers' servants (or batmen) were a potential source of trouble and detailed instructions were imposed on their selection and duties.

> Officers commanding Troops are never to allow a Man to be taken as a Servant whom they think likely to make a Non-Commissioned Officer . . . No Servant is to be allowed to go, or be out of Barracks after watch-setting, unless employed upon his master's business, and with his written leave . . . They will parade once a week in full uniform, with arms and appointments; and on Sunday's Church Parade, will parade with their Troops and fall in on the left of the line.

5 The King's Dragoon Guards in Canada, 1840s

6 The Queen's Bays Warrant Officers' and Sergeants' Mess
India, 1865.

7 The Capture of the Taku Forts, China, 1860.

8 King Cetewayo and his wives being escorted into Sir Garnet Wolseley's camp at Ulundi by his captors, Captain Godson, Captain Gibbins and Lieut Alexander of the King's Dragoon Guards, 31 August, 1879.

The Regulations allowed every officer to have a Dragoon to assist his Servant in the stable. It was insisted that:

> The Officer commanding the Troop, and not the Adjutant, will select the man. Every Second Batman must have his own Troop Horse in the stable with the Officers' horses, where practicable; and he will do all duties as other Soldiers...He must have been three years in the Regiment. It is the duty of Officers to look well after their Second Batman; for, as they are generally good men selected for this purpose ...if they conduct themselves well in it, may lead to their advancement; but which if they fall into habits of idleness and debauchery, too apt to characterize the class of Officers' Servants, as their bad conduct cannot escape observation, must prove most injurious to them.

Determined to try to prevent that other common abuse, Officers misemploying their horses, Colonel Cathcart spelt out the rules plainly:

> All Officers will be provided with a first and a second Charger, approved by the Commanding Officer, and no Officer will part with them without his sanction. No first Charger is on any account to be put in harness, or hacked in any manner liable to render him unfit for service; but as it is desirable that the first Charger should be of the most serviceable description of horse, the permission of the Commanding Officer having been obtained, they may be occasionally used as hunters. No entire horse (stallion) belonging to an Officer will be allowed to be kept in Barracks without the special sanction of the Commanding Officer.

Much of the efficiency of any cavalry regiment depended on the skill and dedication of the Riding Master and his small team of experts. It was enjoined that:

He must set the example of perfect command of temper and patience, both towards Men and Horses and inculcate the same to his Assistants. He must, on no account, send a Man from his Drill to the Adjutant's Horse Drill, until he is perfectly the master of his Horse and has completely acquired the Regimental seat and hand, and is perfect in the elementary movements and formations.

Scales of charges were set out for teaching officers to ride, making their horses fit for the ranks and breaking in their young horses.

Under the Schoolmaster Sergeant, the regiment ran courses for which 'One hour will be appointed every day, at which all Non-Commissioned Officers or Privates, who may not be proficient in writing or accounts, may have the opportunity of improving themselves'. It was stressed that these men should be encouraged to attend. The Schoolmaster Sergeant also looked after the schooling of the children of the married soldiers.

Sixteen Rules were laid down in the Section dealing with Stable Duties. Three rules will be quoted to illustrate the meticulous care and the time spent on keeping the horses in good condition. (They may well have been better treated than the men.)

Rule II. When the weather will not admit of the horses being taken out to exercise, an additional twenty minutes will be devoted to hand-rubbing their legs.

Rule VII. Every horse's feet are to be stuffed with cow dung, at Evening Stables, at least twice a week, or oftener in dry weather, and always the night previous to being shod.

Rule XIV. The present allowance for forage for each horse per day is 10 lbs of oats, 12 lbs of hay, 8 lbs of straw which will be divided into three equal feeds, one to be given at each Stable hour.

Besides the compulsory attendance of all Officers at midday Stables, they were urged to spend an hour or so there daily because it 'will give them the best opportunity of becoming acquainted with the characters and dispositions of their Non-Commissioned Officers and Men.'

A vignette of the progress of the Regiment through the countryside can be glimpsed from these Sections called 'On a March':

> Roads and weather permitting, the rate will be a steady trot of six miles an hour, with an occasional walk, par-ticularly up hills, of about a mile in each hour, and one halt and dismount, to tighten griths and examine shoes, in each five miles . . .After the march the saddles must remain on for at least an hour, or until the horses be cool . . .Staff Sergeants, Officers' Servants and all NCOs and Privates in whatever situation, will invariably march in uniform, and with all appointments . . .Offic-ers will march in red coats and helmets.

Though essential, the baggage was always a hindrance and the Regulations about its position were precise and strict:

> The march of the baggage will be so regulated, that it may arrive at the end of the day's march with the Regiment . . . The Officer or Non-Commissioned Officer in command of the column of baggage must keep the carts well closed up. With this view, he will take care that the slowest and worst appointed carts are in front, and the others not suffered to pass them. No allowance can be made for the baggage of married Sergeants or Men . . .No Soldier's wife, or other person can expect to be allowed to travel on any of the regimental baggage cars.

One wonders how closely these last orders were followed. (See Plate 4.)

Finally, the two matters that caused the gravest concern

were marriage and drunkeness which was very prevalent with so few amenities and spirits so cheap. Standing Orders stated:

> Drunkeness on duty is at all times treated with the severity it deserves, but in parading for, or on a march, it calls for immediate example; and the Articles of War have given power to Commanding Officers to hold Courts Martial for summary example on the line of march.

On marriage the language was even more harsh:

> It is impossible to point out, in too strong terms, the inconveniences that arise, and the evils which follow a Regiment incumbered with women; Officers, therefore, cannot do too much to deter their men from marrying ... No men marrying, not having obtained permission to do so, both from his Troop Commanding Officer and the Commanding Officer of the Regiment, will ever be permitted to receive any of those indulgences bestowed on such as marry by consent.

The section on Marriage concluded on a sombre note:

> The washing of a Troop will assist to provide for a certain number of women – about one for every ten men; but beyond that number great distress must ensue; and it appears only necessary to point out to the men what poverty and misery some of their comrades, who have wives and children, experience to prevent their subjecting themselves to similar hardships.

6

Colonial Wars

UNTIL 1838 neither the KDGs nor the Queen's Bays had campaigned further afield than southern or central Europe, and then only for a very brief period. In the next seventy-five years, not only were they to have long spells of service abroad, but were to serve in every continent except Australia – their duties taking them from China in the East to Canada in the West, and from Edinburgh in the North to Cape Province in South Africa.

In 1837 trouble flared up among the French in Canada. The British Government had become so worried by these risings that a detachment of Regular cavalry was despatched to reinforce the infantry there and, in 1838, three squadrons of the KDGs, accompanied by 270 horses, disembarked at Quebec. Although the first and major part of the rising had been suppressed, there still remained the exhausting, if unspectacular, duty of rounding up scattered rebel bands who posed a threat both near Montreal and around St John's in New Brunswick. In these very remote, rugged and wooded parts of Canada the dissidents could easily slip across the border into the United States, so the cavalry was always on the move, their task being made even more difficult by the onset of winter. After the rising had finally been quelled, with very little loss of life, a tense situation continued and the KDGs were retained in Canada for more than four years. When they returned to the cavalry depot in Maidstone in 1848, the whole Regiment was reunited for the first time in fifteen years. There followed another long spell of duty in the United Kingdom, until 1855 when the Regi-

39

ment, now only 353 strong, was sent out to the Crimea to join the Heavy Cavalry Brigade. During the closing stages of the Siege of Sevastopol, calvalry were not required and the KDGs saw little action and were soon brought home.

The Indian Mutiny broke out on 10 May, 1857, but the news did not reach England until 11 July. On that day the Queen's Bays, then stationed in Dublin, were told that they were to form part of the first reinforcements for India, thus ending nearly half a century of soldiering at home. They crossed to Liverpool and marched from there to Maidstone where they handed over their horses. According to the standard practice of regiments going overseas, the Queen's Bays left behind at the cavalry depot one of their ten troops. Not only were these men to be responsible for helping to train the new recruits who would act as replacements for the casualties, but the home troop would also guarantee the continued existence of the Regiment should the main body be involved in some disaster such as a shipwreck. Although the twenty-eight officers and 635 other ranks could not foresee it, the Bays were to stay in India for over twelve years.

On 25 July, Lieut-Colonel Hylton Brisco and his men boarded the troop transports *Blenheim* and *Monarch*. These vessels were to be their miserable homes for the voyage round the Cape which lasted for over four months. Sailing by the same route two years later, a reinforcement described the appalling conditions:

> She (the *Hanover*) was a two-decker vessel of 1300 tons burthen. The number on board, exclusive of crew, totalled 595 men and 2 females ... our daily rations were 1 lb of biscuits, 12 ozs of meat – always salt meat – never once during the voyage seeing a bit of bread or a bit of fresh meat. We had tea and sugar, of each a small quantity ... Five pints of water per man was issued every morning, of which the cook claimed three pints per man ... The rust from the iron tanks caused the water to become yellow, and it eventually got so bad we had to strain it through a towel before drinking it. The

biscuit was like biting a deal-board. The ship was rationed for a four months' voyage...Each man was served with a three-pound stick of tobacco to last the voyage. Smoking was allowed one hour a day on the forecastle...There was no canteen or other place where necessaries could be obtained...When 28 days out from Gravesend, we sighted Madeira, the only land sighted between England and India. In sight of that island we lay becalmed for 28 days...No boots or shoes were allowed to be worn on board...Medical inspection was carried out daily. To air the messing deck, which served as sleeping room, all hands were turned up on the upper deck. Then one half had to sit down to allow the other half to walk about. After three months at sea...half rations became the order of the day...Every man did his own washing with sea-water for which purpose salt-water soap was supplied, but if you rubbed for a week no sign of lather appeared. The drying ground was over the side of the ship, and on many occasions some villain cut the ropes, allowing all the clothes to drop into the sea.

On arriving at Calcutta late in November, the Bays were informed that they were to go to Allahabad with the utmost speed to join Hope Grant's Cavalry Division in Sir Colin Campbell's force. This meant a 500-mile journey, 400 miles of which entailed marching across India with their newly issued horses. Most of the men had probably never before been out of Britain, let alone seen service in the East, the Bays being one of the first British heavy cavalry Regiments to serve with the East India Company's forces.

From near Calcutta, Captain Seymour, who later became Colonel of the Regiment, wrote:

We are so far capitally mounted...But we shall not get away from this place for another week, notwithstanding we are using almost superhuman efforts to get our saddlery, etc, fitted and ready. Among our horses we

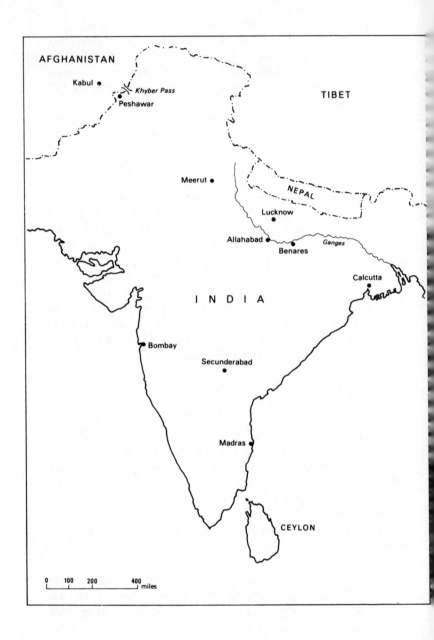

have 120 from the Governor-General's bodyguard, and I believe that we are to get the whole of the remainder.

On the march, he described some of the complications involved in controlling the enormous 'tail' then considered necessary when moving a cavalry regiment.

> Very hard marching we shall have had, as the last ten days are *forced* marches. This, on a *new* regiment, is trying. We are marching with our heavy baggage, yet we have no less than 2,000 *paid* camp followers! We march at 4 am, always in the dark for over two hours, and this adds much to the ordinary confusion of a march. We sleep in tents, but our tents are never up with the Regiment for some time after we reach our new camping ground, being carried on bad elephants and country hackeries. It is all very rough I assure you...Our Officers have been greatly plagued with their native servants, as owing to the times, one is obliged to pay them in advance. Nearly all those engaged in Calcutta have bolted...I fortunately did not engage one there. None of mine have left me...My establishment is complete.

To look after him and his four horses (two chargers and two baggage ponies), Seymour had twelve servants – a head man who was his valet, a cook who acted as butler, a washerman, a tent pitcher, four grooms and four grass cutters!

From outside Benares he complained that out of their 276 dhooley bearers who set out with them all had bolted, but added:

> Our native servants are behaving better than we had reason to expect, though we are quite at their mercy and we are obliged to put up with their roguery, idleness and drunkeness. [He added that the horses were] turning out very well, but it is a nuisance not knowing whether they will stand fire or not – the *sine qua non* of

43

a trooper – as they have never carried a dragoon before. I hear the Carabineer troopers have got the Regiment into several scrapes on this account, having turned tail with their riders.

From the end of January onwards the Bays were regularly employed in minor mopping-up operations which characterized the later stages of the Mutiny and in March they took part in the seige of Lucknow. Later Seymour wrote about an exciting incident:

> We came on bodies of cavalry and infantry of the enemy. 'Bays' ordered to the front to 'charge and pursue!' Away we went as hard as possible, Major Percy Smith and I leading. We did not stop for three miles, cutting down, pursuing and cutting up the pandies[1] right up to Lucknow, and across the river. We are told the most gallant, smartest, though somewhat rash thing that has been done before Lucknow.

This wild dash ended sadly. They were obeying the 'recall' which, in Seymour's words, 'had just been sounding all over the place for us' when a large force of enemy cavalry appeared, killing Major Smith and a corporal. Seymour and the others, however, managed to get back, but he was lucky, his charger having suffered a nasty sabre wound which the veterinary surgeon considered would have finished the animal off it if had been a mere eighth of an inch deeper.

Like all the British units which had to fight through the summer months (it was often over 110 Fahrenheit in their tents), the Bays suffered very heavily, especially from sunstroke. In May, 1858, Seymour wrote, 'We are the healthiest body of troops in or near Lucknow, and yet we have or had over 60 in hospital, out of our strength of 450. Nine of our officers are struck off duty, sick.' In September two soldiers of the Queen's Bays won the newly instituted Victoria Cross. Private Charles Anderson and Trumpeter Thomas Monaghan

---

[1] Colloquial name for a sepoy mutineer.

44

rescued the impetuous Seymour, now the Lieutenant-Colonel of the Regiment (he later became a general) who had been wounded and cut down while attacking some mutineers hiding in the jungle.

After twenty months of campaigning the Regiment ended their nomadic life when, in June, 1859, they handed in their tents to occupy barracks in Lucknow. They had expected to return home in 1861, but were kept on in India as priority was given to those regiments which had served in the Crimean War. In the decade which followed the Mutiny, the Queen's Bays were stationed in several different parts of India, including two tours of duty at Muttra on the North-West Frontier. In 1865, they were honoured by being chosen to provide the personal escort for the Mutiny hero, Lord Lawrence, when he became Viceroy. As was normal in times of peace, their establishment was considerably reduced and in 1867 consisted of 1 colonel, 1 lieutenant-colonel, 2 majors, 7 captains, 7 lieutenants, 7 cornets, 1 paymaster, 1 adjutant, 1 riding-master, 1 quarter-master, 1 surgeon, 2 assistant surgeons, 1 veterinary surgeon, 1 regimental sergeant-major, 7 troop sergeant-majors, 1 quartermaster-sergeant, 1 sergeant-instructor of musketry (a post soon to be abolished), 1 paymaster-sergeant, 1 armourer-sergeant, 1 bandmaster-sergeant, 1 saddler-sergeant, 1 farrier-major, 7 farriers, 1 hospital sergeant, 1 sergeant-instructor of fencing, 1 orderly-room clerk, 1 sergeant-cook, 21 sergeants, 1 trumpet-major, 7 trumpeters, 28 corporals and 378 privates. A little later, the number of farriers was reduced from 7 to 4, but 3 new tradesmen were introduced with 9 shoeing-smiths, 2 saddlers and 1 saddle-tree maker.

When the time came to return home, nearly 100 Other Ranks chose to stay on in India, transferring to other cavalry regiments. This was then a common practice, often for financial reasons, but it can create difficulties when trying to follow the careers of individual soldiers. The Regiment's homeward journey from India was both more comfortable and much quicker than the outward one. It took only one month to go through the Red Sea to Suez. From there they travelled by train to Alexandria and thence by boat to Portsmouth where they landed in January, 1870. They were soon equipped with bay horses and official recognition was now granted to this tradition by a change of title from the 2nd (Queen's) Dragoon Guards to 2nd Dragoon Guards, (Queen's Bays).

The remainder of the nineteenth century was fairly uneventful for the Bays. From 1870 to 1885 they remained in Britain. In 1882, detachments volunteered to serve with other cavalry regiments under Wolseley in Egypt in the Army of Occupation there and a contingent of volunteers joined the Heavy Camel Corps as a part of the abortive expedition to relieve Gordon in Khartoum. In 1885, the Regiment began a ten years' tour of duty in India. The Queen's Bays had achieved and sustained an enviable reputation as a regiment whose men were unusually well disciplined, in the best sense of the word. This may partly be explained by the behaviour of their officers who devoted most of their energies to the well-being of their regiment. In 1895 the Queen's Bays spent a year in Egypt before returning to England where they were at the turn of the century. In 1898, Captain N.M. Smyth, one of their officers attached to Kitchener's Staff, was awarded the VC at the Battle of Omdurman.

Turning again to the history of the KDGs, in 1857 they were also sent to India to help suppress the Mutiny. To their intense frustration, no more horses could then be found capable of carrying men of a 'Heavy' cavalry regiment so they could not be employed on this duty. In 1860 three squadrons (324 men and 270 horses) sailed from Madras to take part in the China War. Its main cause was the refusal of the Chinese to adhere to the 1858 Treaty of Tientsin whereby the British and French were to be permitted to keep Resident Ministers in China. To teach the Chinese a lesson, a joint force of 7,000 French and 14,000 British, under Sir Hope Grant, was assembled and landed on the island of Chusan where a fruitless effort was made to blockade China into submission. When that had obviously failed, the expeditionary force was transferred to the mouth of the Pei-ho river where, on 30 July, it disembarked unopposed. Nearby were the Taku forts and the Chinese had collected a sizeable body of Tartar cavalrymen to bar the way, but these were soon overcome and the advance began towards Peking, 80 miles upstream. Late in September, the Chinese made a final stand near Peking and the cavalry brigade under Brigadier-General Pattle of the KDGs was used in a charge when, in Wolseley's words, 'The King's Dragoon Guards got well in amongst the Tartars, riding over ponies and men, and knocking both down together like so many ninepins. At no time subsequently during the day would they allow our cavalry sufficiently near for a second charge.' Sir Hope Grant's official dispatch for 22 September emphasized 'the excellent services of the King's Dragoon Guards. The country is very unfavourable for cavalry, partly from the number of ditches and partly on account of the crops of maize, etc, having been recently cut, leaving sharp-pointed stubble, most injurious to horses, and rendering their rapid movement very difficult.'

Once Peking had been taken, it was given over to looting and burning. Indeed the behaviour of the Allied troops, especially the French, had been deplorable throughout the campaign. The feelings of the Allies had, however, been aroused by the return of a few surviving prisoners; the rest had died under Chinese torture and included Private Phipps, of the KDGs, who was part of an escort which included some Sikhs. Of him Wolseley wrote, 'Up to the day of his death he never lost heart, and as we were told by one who had been confined with him, always endeavoured to cheer up those about him.' Phipps spoke a little Hindustani which he used to try to keep up the spirits of his fellow-sufferers, mainly Sikhs, who knew no English.

In 1865, the Regiment returned to the United Kingdom where they remained until 1879 when ordered to South Africa. In April, the 634 men and 545 horses of the KDGs arrived at Durban as reinforcements for the second part of the Zulu War. Lord Chelmsford's forces soon began an almost unmolested advance which was marred by an unfortunate incident in which the Prince Imperial of France (Napoleon III's only son) was killed while on reconnaissance; his body was found and brought back by a detachment of the KDGs. In June the Regiment was split up, some being posted to Flying Columns, but most given the less interesting duties of guarding the lines of communication. On 1 July, one officer and 24 men from the Regiment took part in the decisive battle of Ulundi. With the 17th Lancers, the KDGs charged into the retreating Zulu warriors and rode on to the royal kraal, but the warlike King Cetewayo had already fled. Wolseley, who had now superseded Chelmsford, was determined to capture Cetewayo as soon as possible and organized several small columns to search for him. Under Major Barrow, a squadron of the KDGs, with some mounted infantry and native auxiliaries, were soon on the scent and had actually tracked Cetewayo down, but decided to wait till the following day before surrounding his kraal. They were, however, forestalled by Major Marter, also with a squadron of KDGs, who had approached from a different direction and led his men down a precipitous 2,000-foot escarpment to surprise Cetewayo, and the last of the Zulu warrior kings was taken into captivity. In 1880 most of the Regiment sailed for India where they served until 1891.

In South Africa, the KDGs had left behind a small detachment, under Major Brownlow (who later beacme Colonel of the Regiment), who were to return to England to act as depot troops. But before they sailed, the First Boer War broke out and Brownlow was instructed to raise an improvised mounted force in Natal from the remaining KDGs and some of the infantry. The plan was first to reinforce the scattered British garrisons in Transvaal. On 26 January, 1881, the column had reached the Natal-Transvaal border where the road is dominated by Majuba Hill on the left and Laing's Nek on the right. The Boers were visible, digging in to enfilade the approaching British column, but after a very cursory reconnaissance, Brownlow's mounted troops were ordered to clear one of the slopes on Laing's Nek to assist the attacking infantry. They took the wrong route and rashly charged up a very steep hill from which, when they breasted it, they were met with intense fire, Sergeant-Major Lunny of the KDGs being killed instantly and Brownlow's horse being shot under him. Despite some limited initial success, his supporting troop refused to join Brownlow and the whole attack soon had to be called off. The KDGs won their first VC here; the citation reads, 'Private Doogan, servant to Major Brownlow, was charging with his troop, when Major Brownlow's horse was shot. Seeing the Major dismounted among the Boers, he rode up, and though himself severely wounded, dismounted and wished Major Brownlow to take his horse, receiving another wound while trying to get him to take it.'

In the hottest period of the summer of 1882, Francis Younghusband (the famous explorer, leader of the 1904 mission to Tibet, and later a mystic of international renown) joined the KDGs at Meerut on the Indian Plains. This poor, intellectually-inclined, nineteen-year-old officer, burning with military enthusiasm, was well aware that he was a misfit in a smart cavalry regiment. Yet, though critical, his letters home bear witness that he was received by most of his fellow officers with a tolerance and even a sympathy that renders suspect some of the conventional generalizations about the behaviour in Victorian cavalry regiments. (What he writes would also probably be applicable to the Bays.)

> I found, however, that the art of warfare was the last topic of conversation that was likely to rise. There were exceptions of course. The Colonel was a good soldier, and so was Hennah (the adjutant)...I was at first disappointed to find this. I only recovered when I found that these same men, as soon as there was any active service in sight, would move heaven and earth to get there. Keenness for sport did not mean indifference to service in the field. All it meant was disinclination to the monotony of preparation.
>
> On the other hand I was surprised at the friendship

49

I experienced. I had been taught to regard worldliness as synonymous with wickedness, and had expected to find my brother-officers steeped in iniquity. To my surprise I found them excellent fellows; and in my heart of hearts I envied them their good nature. They never went to church except when paraded for service. Their talk was of little else than ponies or dogs. Their language was coarse. And yet they were a cheery lot, always ready to do each other, and even me, a good turn.

After some initial spells of acute depression and of loneliness in which he learned Urdu and Hindustani, Younghusband's biographer records that, 'Best of all were the morning parades. The thought of what the regiment had done in the past and what it might do in the future made me sometimes almost cry with pride . . . to keep one's wits about one and do the right thing at the right time and in the right way – was "to feel a bigger person than one is by oneself alone".' His abilities were quickly recognized and he became adjutant at the early age of 21, but he was soon employed more and more frequently on special duties and afterwards only served infrequently with the Regiment.

To switch from one extreme to another, it is not too far-fetched to suggest that Kipling's poem 'Gentlemen Rankers' may have been based on his knowledge of one of the two regiments, both being in India when this poem was written. It describes the Victorian 'drop-out' who seeks a new life in the cavalry and most regiments had one or two of these men.

Yes, a trooper of the forces who has run his own six horses,
 And faith he went the pace and went it blind,
And the world was more than kin while he held the ready tin,
 But today the Sergeant's something less than kind.
We're poor little lambs who've lost our way,
 Baa! Baa! Baa!
We're little black sheep who've gone astray,
 Baa-aa-aa!

9 The Queen's Bays at Montignies-Les-Lens, 11 November, 1918. From the painting by Lionel Edwards.

10 A KDG Marmon-Harrington armoured car in its 'garage', engaging an enemy aircraft, North Africa, 1941.

11 KDGs escorting Italian PoWs on the 'Axis Highway' near Tobruk, 1941.

Gentlemen-rankers out on the spree,
Damned from here to Eternity,
God ha' mercy on such as we,
Baa! Yah! Bah!

In 1895 His Imperial Majesty Franz Joseph, Emperor of
Austria and King of Hungary, became Colonel-in-Chief of
the King's Dragoon Guards, an appointment terminated
by the outbreak of the First World War. He granted the
Regiment the right to use the Imperial double-headed eagle
as their badge and, except from 1915 to 1938, it has been
worn both by the KDGs and, after the amalgamation, by the
QDGs. The Emperor also ordered that a set of band parts be
sent to the Regiment; this was the Radetzky March, composed by Johann Strauss in 1848 in honour of Field-
Marshall Count Joseph Wenzel Radetzky; it has been the
regimental quick march ever since.

The story behind this unusual link was that both the Czar
and the Kaiser were Colonels-in-Chief of British cavalry
regiments and it was felt fitting that the second longest
reigning monarch (Queen Victoria was the longest) and
occupant of the oldest throne in Europe should be offered
the same honour. (It was also hoped to increase Austrian
goodwill in negotiations with Turkey.) When the Emperor
visited Queen Victoria who was staying in Austria in 1896,
she proposed, on the recommendation of the War Office
and the Prime Minister, that he should become Colonel-in-
Chief of the King's Dragoon Guards. The Emperor was
delighted. Between 1896 and 1914 several officers and men
were invited to stay in Vienna and Franz Joseph took considerable interest in the activities of the Regiment. In the
unlikely event of war between the two countries, he always
promised that if anyone serving in the KDGs was taken prisoner, he would be treated as a personal guest of the
Emperor.

Although the Boer War began in October, 1899, the KDGs
were not sent out to South Africa until early in 1901 and
then they were sent almost straight into action. The majority

51

of the Regiment joined Colonel Bethune's column, but two squadrons were detached to reinforce Plumer's column. These and ten other columns had been collected by Kitchener to try to capture de Wet, the famous Boer guerrilla leader, who had, early in February, crossed the Orange River into Cape Province. In this third, and greatest, de Wet 'hunt', the Regiment was kept at full stretch in the three-week-long pursuit during which de Wet twisted back and forth near the Orange River. On one occasion a small party of KDGs with Plumer's column attacked, unsupported, a rearguard Boer position and were all briefly taken prisoner. On 23 February Lieut-Colonel Mostyn Owen of the KDGs led the advance guard of Plumer's column, consisting of a mixed force of KDGs, Australian and South African horsemen. He covered forty-four miles that day, capturing de Wet's two guns, his ammunition wagons and 102 Boers, but Plumer's men were too exhausted to continue and de Wet, just ahead, escaped that night through a weak and ill-prepared cordon. A lively, but somewhat exaggerated contemporary account of the progress of Bethune's column, in which the majority of the KDGs were serving, can be found in *On the Heels of de Wet* by a 'Staff Officer', the *nom de plume* of Lionel James.

During the rest of the South African war the KDGs took part in several large-scale mobile 'mopping-up' operations, known as 'drives'. In May, 1901, still with Bethune, they were engaged in a sweep of the north-eastern Orange Free State; in August, now with Broadwood's column, they were in a large 'drive' across the western part of the Orange Free State, and at the end of the year they formed part of de Lisle's column, again in north-eastern Orange Free State. Like all the horsed formations, the KDGs were continually on the move, covering great distances pursuing an elusive enemy with meagre results. With simpler and lighter equipment, and often with two ponies, one of which carried their kit, the Boers could nearly always outpace and often outfight the cavalryman whose horse had to carry a much heavier load.

C.J. Briggs made his reputation in this campaign. He went out on the Staff and was wounded in the Battle of Magersfontein in January, 1900. He was later attached to the Imperial Light Horse, the most prestigious Uitlander (British South African) Regiment, which he commanded from September, 1901. He specialized in leading night attacks against the Boer commandos, a type of operation pioneered by Colonel Benson. Altogether the KDGs lost four officers and eight other ranks killed, twenty-seven, all other ranks, from disease (mainly typhoid) and thus, with the wounded, had a total casualty list of fourteen officers and sixty-nine other ranks. The Regiment stayed in South Africa for a year after the end of the war and were then posted to England from 1903 to 1908 before going to India where they were at the outbreak of the First World War in 1914.

The Bays were kept in England for most of the Boer War, but their strength was increased to over 900 all ranks and they provided several drafts for South Africa. On 2 February, 1900, 100 men from the Regiment, led by Lieut-Colonel Dewar, marched in front of the gun carriage carrying the coffin at Queen Victoria's funeral.

In December, 1901, 537 officers and men, with 488 of their own horses, landed at Cape Town. They were posted to Colonel Lawley's column which, after de Wet's flight into Transvaal, rounded up some of his followers in the northern part of Orange Free State. On 31 March, 1902, Lieut-Colonel Fanshawe, their Commanding Officer, had been informed that two small commandos, each estimated to be 200 strong, were laagered about twelve miles away near Leeuwkop in a hilly region some forty miles east of Johannesburg. Lawley ordered Fanshawe to make a surprise night attack under the guidance of a British intelligence officer and forty National Scouts, (Boers fighting on the British side), who were attached to the Regiment. The attack succeeded and at 0330 hours 284 Bays surprised the laager, taking several eminent prisoners, including Commandant Pretorius whose father had given his name to the capital of the Transvaal. Fanshawe decided to attack the other laager, but, because of inaccurate intelligence, the Bays now found themselves involved not with 500 or so enemy, but with 800-1,000 experienced Boer guerrillas who knew the country well and were in a most aggressive temper. Fanshawe had therefore to organize a fighting withdrawal by squadrons twelve miles back to his base where he could expect help from the column. With the Boers all around them, the Bays fought desperately as they fell back, losing more men each time. The Boer attacks only ceased after daylight when the Bays had reached the cover of the column's guns and the other cavalry regiment had moved out to help them. In this sharp running fight the Bays had two Officers and twenty-one other ranks killed and nearly fifty wounded, as well as 120 horses killed or badly wounded. The Bays had, however, inflicted an unusually large number of casualties on the Boers who had had at least five of their senior men killed and probably over forty casualties all told. This very minor battle of Leeuwkop, or Boschman's Kop, was typical of the Boer War and showed how hard it was ever to obtain reliable information about Boer strengths; what had happened was that the Bays had stumbled on a meeting of several commandos hastily called together both to discuss the peace proposals and, if possible, to inflict a setback on Lawley's Column.

The Bays were kept very busy till the end of the war (31 May), having marched 900 miles over the veldt between 8 April and 10 May. In this campaign they suffered a total of 100 casualties, nearly all at Leeuwkop, but nineteen also died from disease; the horses lost reached the incredible total of 748, out of 775 with which they had started the war. The Bays remained with the Army of Occupation until 1908 when they took over the Hounslow Barracks from the KDGs.

7

The Eclipse of the Horse

WHEN the First World War broke out the Bays were in
Aldershot with the 1st Cavalry Brigade commanded by
Brigadier C.J. Briggs of the KDG. On 5 August mobilization
started and was completed in five days, bringing their numbers up to 660 all ranks. Many of the Reservists who were
recalled had served in South Africa. These experienced men
formed 36% of the total other rank strength and were an
invaluable addition to the Regiment which was largely composed of young soldiers. The Bays had their full quota of
officers, which included three from other regiments, two
being from the KDGs who were in India. On 11 August King
George V and Queen Mary inspected the Bays, who sailed
for France on 16 August as part of Allenby's 1st Cavalry
Division in the British Expeditionary Force. By 19 August
the Regiment had concentrated near Maubeuge, about
twenty miles south of Mons, where they were then the most
advanced of the British units.

The 70,000 men of the BEF were commanded by Field-
Marshal Sir John French and consisted of I Corps under
Haig and II Corps under Smith-Dorrien. By 22 August
Haig's two divisions had been positioned to the east of Mons
and Smith-Dorrien's to the west. The Cavalry Division was
held in reserve with most of its formations behind II Corps;
the Bays were billeted in the village of Andregnies where the
nuns had prepared a large supper in their girls' school. Here
the officers were to pass their last peaceful night for many
weeks. On 23 August the four Corps and three cavalry divisions of von Kluck's 1st Army (about 160,000 strong) clashed

with the British in the day-long battle of Mons. The French armies were falling back everywhere and, if they were not to be cut off, the British had to conform. Very late that night the Bays received orders to begin a retreat that lasted eleven days as the BEF fell back south of the River Marne, covering a distance of over 150 miles.

On the first day of this famous retreat the Bays acted as one of the rearguard regiments to protect the infantry and had only a minor brush with the Germans. They were not involved in the delaying battle at Le Cateau, but resumed their rearguard role as the retreat continued through the night of 26 August. Later the following day some of the Regiment reached St Quentin and one of them recalled, 'I, personally, hardly knew we were in the town at all, being fast asleep on my horse.' The roads were jammed with the marching troops, guns, wagons and refugees which compelled the cavalry to try to keep on parallel lines with the main body by finding their way through adjoining fields and by skirting round the villages en route. Such a laborious and wearisome manner of moving caused considerable dispersion, but Lieut-Colonel Wilberforce gathered most of the Bays together in St Quentin, except for 'B' Squadron which had become detached. Nevertheless Major Ing, Comanding 'B' Squadron, pressed on, collecting numerous stragglers which almost doubled the strength of his squadron, before rejoining the regiment on 30 August.

On that day 1st Cavalry Brigade reached the small village of Nery, just south of the Forest of Compiégne. Here they passed the night and, being in reserve, assumed they were secure, but von Kluck's army had unexpectedly swung eastwards to try to cut off Lanrezac's Army; the Germans considered that the BEF had already been destroyed. Early that morning one of the German cavalry divisions surprised the 1st British Cavalry Brigade. Among their successes the German guns hit the Bays' horse lines causing many of the animals to stampede, but Lieutenant Lamb managed to bring his two machine guns into action. With seven men, he fired thousands of rounds, aiming at the flashes of the German guns, and gave some much needed support to 'L' Battery, RHA, who were virtually wiped out serving their guns. Meanwhile, Brigadier Briggs had reacted quickly to this critical situation and despatched the Bays and other units in the Brigade to take up defensive positions. Their rifle fire, Lamb's machine guns, and the artillery held up the German division for two hours until reinforcements arrived. In the British counter-attack, the Germans withdrew, leaving many killed, and eight of their twelve guns abandoned, as well as losing seventy-eight prisoners. Altogether about 110 of the Bays had fought at Nery; they lost five killed and forty-three wounded. Lieutenant Lamb was awarded the DSO and Private Ellicock the DCM for their part in this action.

Later that morning the retreat was resumed, but although the great heat made conditions extremely uncomfortable, the Bays were engaged in no more fighting. On 4 September they crossed the Marne about seven miles east of Paris and two days later they reached the southernmost limit of their withdrawal. Although a number of their men had rejoined the Regiment, they were seriously under strength, having lost seven of their twenty-four officers, and 113 of their other ranks; their losses in horses were extremely heavy, 223 of their 527 riding horses and twenty-six of their seventy-two draught horses having been lost.

On 6 September the tide began to turn as the Allies advanced 50 miles northwards towards the River Aisne. On 12 September, the Bays came up against a German strong-point at Braisne a few miles south of the river. This village was strongly fortified and its capture demanded unpleasant street-fighting, causing the Regiment some casualties, but they soon cleared out the Germans and reached the banks of the Aisne, just beyond which there were to be terrible battles. On 3 October they were ordered about 150 miles northward to the Béthune-La-Bassée area where they arrived on 11 October, their march having taken them parallel with the newly emerging front line. Most of the British forces were now being concentrated on the Franco-Belgian border where a race to the sea was in progress with the

Germans. On 30 and 31 October the Bays were in some very fierce defensive fighting at Messines. During November, in the First Battle of Ypres, the 1st Cavalry Division was in reserve and they were continually being sent from one trench system to another to plug gaps.

Trench warfare had now begun in earnest and the use of cavalry as cavalry became increasingly improbable. In common with all the the other mounted regiments, the Bays were therefore regularly employed in infantry duties, often in an emergency to help stop a breakthrough by the enemy, as happened during the latter stages of the Second Battle of Ypres in the spring of 1915. Also they had to provide working parties for maintenance tasks near the front line. One of their officers explained how unnatural this transition was:

> The cavalry of all arms had the hardest work to adapt themselves to this constricted method of fighting. All the training, both of officers and men, taught one to look away into the distance, to think over the hill and to use the mobility that our association with the horse gave us. Now this was all changed: one pored over maps not of miles to an inch, but of inches to a mile.

This kind of dreary, expensive and horrible soldiering was to last almost uninterruptedly until August, 1918.

From 1915 to 1917 the Bays were mostly kept in reserve. It was always confidently anticipated that each of the great offensives would result in a massive breakthrough which the cavalry would then exploit. In 1915 at Loos, in 1916 at the Somme (where it was claimed that the biggest concentration of horsemen in Europe for a century had been assembled) and in 1917 at Arras and then at Cambrai, very large forces of cavalry were concentrated behind the lines and waited impatiently for the great moment which never materialized.

In March, 1918, when the German offensive threatened to overwhelm Gough's Fifth Army, the Bays helped to stem the enemy's advance and endured some of the fiercest fighting of the whole war. In fourteen days they suffered 155 casualties, over a quarter of their effective strength; this was nearly twice their total losses for 1917. Even in the final battles of August and September when the German defences began to crumble, the cavalry could not be freely employed, mainly because modern artillery and machine guns were so lethal against horses. On Armistice Day the Bays, still with 1st Cavalry Division, were in Belgium at Montignies-Les-Lens, about seven miles north of Mons and only a few miles from where they had started the war. On 1 December they marched into Germany and were billeted near Cologne. In the First World War the Bays' total casualties were 577 and of this total, eleven officers and 143 other ranks had been killed or died of wounds, and thirty-four other ranks were missing.

In November, 1914, the KDGs landed in Marseilles as part of the Indian Expeditionary Force. They brought with them 604 horses, the great majority of these being Indian country-breds, animals emanating from the Army's own studs. The KDGs' experiences in France were now very similar to those of the Bays. Their first spell in the trenches was in January, 1915, near Festubert. The conditions there were appalling with the water coming up to the men's armpits and the Regiment had to evacuate eighty-two men suffering from exposure. In June, as part of the Lucknow Brigade in the Indian Cavalry Corps, the KDGs fought their most memorable battle at Hooge Château near Ypres. This building had been fortified as a strongpoint, but, after being shelled for nearly two days, which destroyed all except two walls and the stables, the Germans attacked. Only four of the defenders managed to return and three of them had been wounded. Another squadron of the KDGs quickly counterattacked and regained the stables. In this relatively minor action, the losses were thirty-four killed and fifty wounded. General Plumer, commanding 2nd Army, sent a personal message of appreciation to Major Turner, the acting Commanding Officer, for the Regiment's gallantry at Hooge.

In 1915 the Regiment moved thirty-five times, in 1916

twenty-five times and in 1917 twenty-two times and these moves included nine tours of duty in the trenches of an average duration of ten days. As with the Bays, most of these years were passed in waiting to exploit the great breakthrough which never materialized. The KDGs' casualties were six officers and eight-two other ranks killed (twenty-five of whom were attached to the Life Guards) and 166 of all ranks wounded. In October, 1917, the Regiment was brought up to strength with thirty-one officers, 566 other ranks and 585 horses and they sailed from Marseilles for India, arriving there in November.

In May, 1919, the KDGs took part in the little known Third Afghan War, the first time the Regiment had been in action on the Indian sub-continent. A rising occurred among the Afghans in the city of Peshawar and the Regiment was ordered to make a thirty-mile night march to close its sixteen gates and prevent the troublemakers from escaping. This was successfully accomplished and the investment of Peshawar was handed over to the infantry and police, the revolt soon collapsing. A few days later, as a part of a composite group that included three cavalry regiments, the KDGs advanced through the Khyber Pass into Afghanistan and captured the village of Dakka just off the main Kabul road. Two days later, three squadrons went forward with a small reconnaissance party which pursued some Afghan tribesmen to the far side of a nearby mountain pass. The British force now found itself under quite heavy fire and this attracted the attention of a much larger body of Afghans who were concentrating to march on Dakka to try and recapture it. Greatly outnumbered, the British began to withdraw and the Afghans, observing this, attacked with increasing determination, causing losses both to the cavalry and the infantry. Placing their own and some of the other wounded on horses, the KDGs helped extricate the force through a pass, but a stretch of open ground still lay between them and the village where those in the main group had not realized their plight. The commander of the reconnaissance force asked Captain Cooper, in charge of the KDG detachment, if he could charge the Afghans. Cooper agreed and led the charge, scattering the enemy and enabling the British to rejoin the rest of the group unmolested. This was almost certainly the last cavalry charge made by the British Army. Captain Cooper was among the twenty-two KDGs wounded in ths action in which one officer and six men were also killed; Cooper was later awarded the DSO, another officer the MC and two of the men the DCM. The composite group was besieged for nearly two days at Dakka before being relieved by the main expeditionary force. During the remainder of this war, which ended in August, the KDGs incurred a few more casulaties.

In January, 1920, the KDGs were on the move again, this time going to Basra in Mesopotamia (now Iraq). Due to demobilization, the Regiment had become very depleted and mustered only eleven officers and 140 other ranks, but their numbers were soon increased considerably. In the aftermath of the collapse of the Turkish Empire, the Regiment was kept busy on peace-keeping duties in Iraq and for a short time in Persia. In 1921 they came back briefly to Edinburgh before joining the Army of the Rhine for three years. This was followed by eight years in England, before leaving in 1932 for Egypt where they remained until 1935 when they began their last tour of duty in India, staying there until 1937, when the Regiment was ordered home to be mechanized. Their final ceremonial mounted parade was held at Secunderabad. With Lieut-Colonel Tiarks at their head, the three long lines of pith-helmeted soldiers were drawn up, swords unsheathed, on their shiny black horses. This was perhaps one of the most moving events in the two and a half centuries of the KDGs' history.

It may be recalled that over 600 Bays had sailed from Southampton for France at the outbreak of the First World War. At the end of March, 1919, a mere sixty disembarked at the same port. After the casualties and demobilization had taken their toll, they were all that was left of the Regiment. They only stayed three months in England, training new recruits, before they were off again, with their strength

61

increased to nineteen officers and 499 other ranks. Their first stop was El Kantara in Egypt where a huge remount depot contained 50,000 horses; having acquired their mounts, their destination was Beirut where they found themselves employed in keeping the peace in Syria. When the French arrived in November the 12th Cavalry Brigade, headed by the Bays, set out on a 500-mile march, from Aleppo to Gaza, which took four weeks, during three of which it rained continuously. For most of 1920 they were stationed in Palestine and sailed at the end of the year for India where they remained for seven years. In 1921 it was announced that the Regiment would henceforth be known as 'The Queen Bays' (2nd Dragoon Guards). In 1935 the Bays were to lose their horses and be mechanized.

In the last days of the horsed cavalry the establishment of a regiment was Regimental Headquarters, HQ Squadron, Machine Gun Squadron and three Sabre Squadrons, each of three troops. There were thirty-two officers, 566 other ranks and 585 horses. All carried swords, but the officers and senior NCOs were armed with revolvers not rifles.

A cavalry regiment had three main roles: Reconnaissance, Pursuit, either surprise attack mounted or outflanking the enemy, and Delaying Action which was performed dismounted, using machine guns and rifles. Weapons were not fired from the saddle. Except for very brief periods or to surmount obstacles, galloping was unusual. This was because the horse, carrying 250-300 lbs, was too heavily laden to be able to charge across country. In addition to its rider, a blanket and a saddle blanket were carried. Fitted to the front arch of the saddle was the ground sheet and two wallets with a change of clothing and washing kit, and on the back arch a ground sheet and greatcoat as well as feeds of 3½ lbs slung on either side. On the near side of the saddle was the shoe case containing a front and a rear shoe (already made to fit the horse) and nails. Outside this was the sword in its scabbard, the surcingle pad and grooming kit. On the offside was the rifle in its bucket, a shackle peg, mess tin and a canvas water bucket. The horse's headdress consisted of a bridle and head collar with a head rope and a 'built up' rope (loop at one end and toggle at the other). Joined together these formed a ground line; where possible, however, a breast line was rigged up in camp.

Between the two World Wars 1st Cavalry Brigade was stationed at Aldershot and 2nd at Tidworth. A regiment usually spent three years with a brigade and one as a Divisional cavalry regiment. The Divisional cavalry regiments were stationed at Edinburgh, York, Colchester, Shorncliffe and Hounslow from where regiments began their foreign tours of two years in Egypt and five in India. Those in India were kept up to strength by drafts of recruits trained by the home regiments, which put a considerable strain on their personnel. The changing of stations and sending drafts overseas was always done in the trooping season from October to March.

In 1926 the Cavalry Depot at Canterbury was closed and all training was undertaken in the regiment. Provided they passed the medical tests recruits were accepted, so the standards of intelligence varied considerably; quite a few, being unable to read or write, had to be given basic education. The first stage of training consisted of musketry and foot drill, both to improve their physical fitness and to introduce recruits to the kind of manoeuvre performed on horseback. Thus when the order was given, 'Sections (or troop) about wheel' the front rank advanced four paces before beginning the wheel – corresponding to the advance of a horse's length. This was to enable the wheel to take place without horses in the rear ranks treading on the heels of those in front. After passing out on foot drill, recruits were posted to Squadrons to begin Riding School on horses from the Squadron lines. This lasted about one year, the average recruit taking 90-100 lessons of one and a half hours each. It was a progressive course and a recruit could not move on until he had mastered the previous stage. Some of the best cavalrymen were miners, among whom recruiting was good because of unemployment. The Riding School was the responsibility of the Equitation Officer, whose staff consisted of a Riding Sergeant-Major, a Sergeant, a Corporal and a Lance Corporal. Good riders were frequently used as auxiliary instructors. The aim was to keep each ride to no more than twelve men. Before being posted to their Troops, all recruits had to be passed out by the Squadron Leader and Commanding Officer on riding and all other aspects of their training. A newly arrived officer also had to be passed out on musketry, drill and equitation by the CO and was taught how to shoe and groom a horse, as were the Troopers. The only variation in the training routine abroad was that it took place in the cooler winter months. In Britain individual training lasted from January to the end of March – everyone had then to revise their basic skills. Whatever job a man might be doing, he was a cavalryman first, and thus Riding School had to be performed all over again; besides training the men, it put the horses through their paces and the more experienced riders were allotted remounts to train. Senior NCOs were made into what was called the Staff Ride, usually performed under the eyes of the RSM or the Riding Sergeant-Major. All other ranks had also to requalify in their respective jobs. The next phase was troop training, when the tactical duties of a troop were practised under the Squadron Leaders. Particularly important was the handling of the led horses. No. 3 in each section was the horseholder, who was responsible for controlling the horses when their riders were dismounted. Trumpet calls were invariably used to signal drill movements and changes in pace as they were understood by the horses. The Commanding Officer, the Second-in-Command, the Adjutant and each Squadron Leader had a Trumpeter riding at his side. The jingle and the dispersion of a horsed regiment made verbal commands impractical. Some of the less musical soldiers found it easier to identify the calls by fitting rude words to them, such as, for the Trot 'arseholes bobbing up and down'; Advance sounded like 'come along, come along', and so on. The object also was to get the horses fit enough to cover up to 20 miles a day for a long period.

Then followed Squadron training under the supervision of

the CO, and finally Regimental and Brigade training, in areas such as Salisbury Plain. At the end of every day's training or work, the Troop Leader, or Troop Sergeant, inspected every horse to check that it was fit for work the next day. He always remained on duty until every horse had been groomed, fed and bedded down for the night. The Troop Farrier was always on hand to replace lost, or refix loose, shoes, and the Saddler to repair broken pieces of saddlery. In Barracks there was a Regimental Forge under the Farrier Sergeant-Major, in which farriers, including the Squadron Farrier Sergeant and Corporal worked together.

If a horse was sick or injured it was normally brought from the Squadron to the Forge where, if necessary, the Farrier Sergeant-Major arranged for the Veterinary Officer to inspect it. Every regiment had attached a Royal Army Veterinary Corps officer, usually a major. The CO made a regular inspection of all the horses, accompanied by the Veterinary Officer. If a horse was too sick to be cared for by the regiment, it was sent to the Station Veterinary Hospital, run by the RAVC, until it was fit to return to duty. If it had to be destroyed, it was despatched at the hospital; each year a number of horses were 'cast' and destroyed on account of age.

The horses were either geldings or mares and their average age was about nine years. Every horse had its Army number tattooed on its upper gum, and to read it the top lip had to be raised. This practice was discontinued in the early 1930's when the stealing of army horses became unknown. On one hoof was branded the Regiment's identity (e.g. KDG), on another the Squadron number (e.g. A 137). If necessary, these were renewed when the animal was shod. Three years was the longest period a regiment kept the same horses, as they were handed over when changing stations, the regiment moving, not the horses, except when on active service.

A regiment had no choice over its horses, all coming from the Army remount depots. Most of them were bought as four-year-olds from Ireland; in India they were mainly walers

65

from New South Wales. Every year ship-loads of remounts were transported from Australia to Calcutta, Madras and Bombay.

Every officer had two chargers, also from the remount depot, and these were kept in separate stables and looked after by his batman (known as the First Servant). In the leave season (October-December) an officer could hire two horses from the troop for hunting; these were called 'fifteen bobbers' as this was the official charge and a trooper (Second Servant) looked after these. With one-third of the personnel away, the leave period was a particularly busy one. To prevent the horses becoming too fresh and unmanageable, they were exercised for two hours every day (8.30-10.30) by a rider with one horse on either side of him; if there was snow and ice outside, stable manure was put in a circle on the barrack square, and the two hours exercise was spent going round and round this track.

Sickness and various essential duties meant that the majority of the troopers and junior NCOs had two horses under their care. In winter the horses were clipped out to prevent them sweating. Cavalry horses had no mane and thus their necks needed clipping – known as 'hogging' – at frequent intervals. Considerable time and effort was spent in shaping the tails, by pulling them (i.e. removing individual hairs), straggly ones being singed by a spirit lamp, a task usually performed by the Troop Sergeant. The ends of the tails were 'banged' or cut at hock level with a special guillotine device by the Squadron Leader or Second-in-Command. The shaping of the tail was also improved by a tail bandage, often made from an old puttee and generally worn in stables. A newly joined officer soon learnt to recognize every horse by its tail. The story goes that one Squadron Leader wanted all his horses to keep their tails up for a special occasion and ordered that a stick of ginger should be inserted up their rectums, but, an officer recalled, 'the result was disastrous because as they walked past it did not seem natural'.

The horses were never left unattended. By day, each Troop had a stable picquet from Reveille until Guard

12 The Prime Minister with Lieut-General Martel, General Sikorski, General de Gaulle and Major-General Noons visiting the Queen's Bays prior to their departure for North Africa, February, 1941.

13 Queen's Bays tank crews brewing up beside their 'General Grant' tanks in the desert, 1941.

14 The Colonel-in-Chief presents the new Standard to the
Queen's Dragoon Guards at Clarence House, March, 1959.

Mounting when there was a roving picquet who was on duty around the stables all night. The Orderly Officer used to go round with the Sergeant of the Guard some time after 2300 hours every night to check that everything was in order, in case, say, a horse slipped its head-collar and got loose. Wives and families were not allowed in the stables area, except on Sundays in Britain, when the stables were open to the public after church parade and large numbers used to walk round to see the horses bedded down.

A horse was allowed 8 lbs of oats, 10 lbs of hay and 10 lbs of straw per day. This was indented for by the Quartermaster and issued to the Squadron, each of whom had a Forage Orderly, usually a lance-corporal, who in turn issued it to the troops and to the officers for their chargers. Under the supervision of the Squadron Leader, thinner horses were fed up at the expense of fatter ones and every troop kept a forage roll, showing exactly how much each horse was entitled to. The Troop Sergeant normally checked that the individual rations (agreed on in consultation with the Troop Officer) had been correctly weighed out for the feeds and that the oats had been properly crushed, the crusher being in the regimental forage barn. On Saturdays some linseed was added to the evening feed.

The Troop Sergeants were always supposed to pick the quietest and fattest horses for their own mounts! If there was a particularly good horse, a jumper for instance, it was given to the best rider in the troop. The officers, many of whom had been riding since they were small boys, were, on the whole, better riders than the men. They had also been trained at Sandhurst for two years where three periods of 1½ hours weekly were devoted to riding instruction. Many officers also played polo regularly two or three times a week.

The relationship between the officers and men was clearly defined. Each was expected to keep on his own side of the fence, although if a trooper had a personal problem he would usually prefer to discuss it with his Troop Officer than with anyone else. The only mixing that occurred was at regimental functions such as dances; and annually on Waterloo Day the KDG officers dined as guests in the Sergeants' Mess.

Finally, for the best riders, NCOs and officers, there was the much cherished opportunity of being sent on the year's Instructors Course at Weedon. Here everyone was allotted three horses. The trained one was used for the training in the usual riding school methods to the most rigorous standards, the half-trained horse was a remount from the previous course and had to be fully trained by the end of the course and finally a raw, untrained remount had to be half-trained. In addition to being taught advanced equitation, a considerable amount of veterinary knowledge and military training theory were included in the syllabus. An enjoyable part of the training was three days' hunting a week with the local packs, the Grafton, Pytchley and Warwickshire. Practical and written examinations were held at the end of the course. From those who had attended the Weedon course were chosen the Riding Master and Riding Sergeant-Major in the Regiment. For over a century the daily routine of a cavalry regiment was generally as follows:-

Reveille	0530	(In winter 0600)
Roll call	0545	This was followed by mucking out and watering and feeding the horses.
Breakfast	0730	Between early morning stables and breakfast ablutions were performed.
First Parade	0830	This could cover all phases of training and usually lasted until 1030 when the horses were watered and fed.
General Stables	1100 1230	During this period every horse was groomed, and inspected by

		the Troop Officer and it was during this period that the Squadron Leader visited the stables. Horses watered and fed.
Lunch	1300	
	1400	Riding School, education, clean-
	1630	ing of tack and many other general duties were performed.
Evening Stables	1645	Horses were usually given a further rub down, bedded down and fed. Rugs were put on in winter.
Evening meal	1730	
Late Water	1900	This was done on a section basis and in summer hay nets were put up.
Last Post	2000	
Lights Out	2015	

Although this routine may now appear very demanding, one of those who served in this period remarked, 'Believe me it was a good life and enjoyed by most'.

8

Tanks and Armoured Cars — France and North Africa

Exactly a quarter of a millenium after their foundation, the Queen's Bays exchanged their horses for tanks. To achieve such a complete transformation in their way of life was a difficult and slow process as all those serving needed to be retrained. Thus from 1935 to 1938 the Regiment largely became a training school with many men always away on specialist courses, while within the Bays itself a large number of courses had to be run to accustom the officers and soldiers to their new mechanized equipment. In addition, nearly all the stores had to be replaced by the complex parts needed to keep the new vehicles and guns in operation. Inevitably some of the older men found the transition too difficult and, where possible, had to be given other jobs. The Regiment's task was made more frustrating and onerous by the acute shortage of tanks which afflicted the British Army. Nevertheless, by the beginning of 1939, the Bays were fully equipped with light tanks and joined 2nd Armoured Brigade (with the 9th Lancers and the 10th Hussars) in which they fought for most of the Second World War. In July, 1939, reservists were called up, and those posted in included soldiers who had seen service in other tank regiments; they brought the Bays up to full strength for the first time since 1927.

The winter of 1939/40 was spent in further training and receiving new equipment. This included the A.9 cruiser tank mounting a two-pounder gun. Early in May the Regiment was ordered to France as part of 1st British Armoured Division and was only then issued with some of the newer marks of cruiser tanks, the A.10 and A.13. The intention was that the Division would finish its training in France. Before sailing the Bays were inspected by Queen Elizabeth, who was their Colonel-in-Chief, and by King George VI.

The Bays reached Cherbourg on 20 May and their twenty-nine cruiser and twenty-one Mark VIC light tanks were put on trains and two days later were unloaded at the small station of Bréval, about half way between Rouen and Paris, south of the Seine. The German offensive had begun on 16 May and by 22 May their forces had reached the coast at Boulogne and Calais, thus cutting off the newly arrived British formations from their main army. Confusion reigned, but the plan was for the British armour to cross the Seine and hold the Germans on the Somme near Amiens where, on 23 May, the Bays went into action for the first time as a tank unit, but without proper infantry support. They fought bravely in this area until 7 June, trying to help the 51st Highland Division. The cruiser tanks were proving unreliable, their brake-linings having burnt out (no spares were available), which made them almost unsteerable and put them at a grave disadvantage against the Germans who now began their offensive in this area. On 8 June the 1st Armoured Division were moved south of the Seine and ordered to blow up all the bridges round Rouen and to hold the river line until the French arrived to help them. As French resistance collapsed, 1st Armoured Division, now almost without any tanks, pulled back westwards to Le Mans where, on 15 June, it was heard that the French had asked for an armistice. Later that day most of the Regiment was ordered to take their lorries to Brest, 240 miles away, and embark at once for England; they reached Plymouth on 17 June. The few tank crews who had been separated from the rest of the Regiment left from Cherbourg on 18 June. Although they had lost all their equipment, the Bays had had an almost miraculous escape from captivity and had only lost two officers and twelve men. They remained in England as a part of the defence forces until September, 1941, when they were again fully equipped as an armoured regiment.

At the end of 1937, the KDGs returned from India and almost immediately began their conversion from a horsed to a mechanized regiment. This process was just about completed when the Second World War broke out and the Regiment, then in Aldershot, was inundated with 1,400 reservists, its task being to receive them before posting on most of them to other regiments. By mid-September, when it moved to Dorset, the KDGs' strength was 26 officers and 480 other ranks, this number being increased to 537 by the end of 1939. The Regiment was equipped with light tanks (Mark VIB) which mounted only two machine guns and many of these were transferred to another regiment in the spring of 1940. Nevertheless directly after Dunkirk, the KDGs, with twenty or so tanks, became one of the few armoured units in Britain and thus an important element in the small anti-invasion force. After a spell in Yorkshire, the Regiment was moved to Newmarket where it formed a part of the Mobile Reserve Force responsible for the protection of the shores of southern and eastern England, should the Germans make a landing.

During November an advance party, and a little later, the main body of the Regiment sailed in different convoys round the Cape for Egypt where they were reunited on 31 December, 1940. The following day the CO was told that they were to be converted into an armoured car regiment and for a cavalry regiment it was a far more attractive role, fitting much better into their traditional pattern of warfare. The main reason why the KDGs had been chosen was that they had served in Egypt from 1932 to 1935 and many of the officers and men knew the Western Desert. The Regiment was equipped with the South African-built Marmon Harrington, a new type of armoured car. It was a make-shift machine, being very tall, thinly armoured and armed merely with a light machine gun and the feeble Boyes anti-tank rifle; a medium machine gun was provided as anti-aircraft protection. The main virtue of this unknown machine proved to be the mechanical reliablity of its standard 30 hp V.8 Ford engine. Later the Marmon Harringtons were greatly improved by removing the turret and fixing on a two-pounder or a captured German or Italian gun of similar calibre, but this modification was highly unofficial and had to be removed before sending cars back for repairs.

On 27 January, 1941, Major Crossley set out from near Cairo with fourteen armoured cars of 'B' Squadron. The first stage of the journey took four days and they covered about 300 miles along the rough coastal road to join the rear formations of 7th Armoured Division at Bardia; they then moved on another 150 miles westward to Mechili, a junction of many desert tracks. With the recent capture of the fort at Tobruk, Wavell decided to support O'Connor's daring plan to by-pass the Cyrenaican hump and try to cut off the rest of the Italian Army which was retreating into Tripolitania. Starting from Mechili on 4 February a small advance guard was to plunge 150 miles, mainly across unmapped desert, to cut off the retreating Italians on the coast south of Benghazi. The armoured car group consisted of RHQ and one squadron of 11th Hussars (7th Armoured Division's now depleted armoured car regiment) and 'B' Squadron, KDGs, which was attached to them. This famous advance culminated in the Battle of Beda Fomm (5-7 February) in which 'B' Squadron played an important part in the cutting of the coastal road and preventing the retreating Italians from breaking through; the Squadron collected over 500 prisoners. Soon afterwards 7th Armoured Division was pulled back to refit and the 11th Hussars handed over to the KDGs at El Agheila, the remainder of the Regiment having just arrived. Thus, after seven weeks as an armoured car regiment, they found themselves holding the most forward position in the Middle East Army, on the Tripolitanian border, where they remained until the end of March.

On 20 February a KDG officer reported almost having collided with an eight-wheeled German armoured car near Agheila. This, the first report of Rommel's Afrika Korps, was ignored by all the Headquarters in the rear. Within a few days, however, the growing activity of the Luftwaffe as well as the appearance of German tanks began to place a serious strain on the weak British units stationed near the Tripolitanian-Cyrenaican border. As the main British commitment was now to assist Greece, few reinforcements were available and authority was given for a withdrawal, if necessary, to Benghazi about 150 miles to the north. The strength and offensive skill of Rommel's forces was seriously underestimated and the withdrawal soon turned into a full-scale retreat which began on 1 April. A week later the KDGs found themselves in Tobruk, having kept together, despite some losses, during a chaotic journey of 400 miles in which almost every senior commander had been captured, including General O'Connor, the main architect of the very recent victory over the Italian Army. The Regimental History summarized the prevailing mood when it said, 'All we were conscious of was a feeling of bewilderment and frustration; for weeks the Regiment's patrols had given warning about the presence of a substantial German armoured force, yet the attack seemed to have taken the Western Desert force completely by surprise'.

In this crisis Wavell had flown up from Cairo and personally ordered that the fortress of Tobruk was to be held at all costs. In fact, the fortuitously assembled garrison resisted a siege of over seven months (11 April-1 December, 1941) and their heroism gripped the imagination of the British public more than almost any other event in the Second World War. Although the main formation was the newly arrived 9th Australian Infantry Division, it had no supporting units and, throughout the Siege of Tobruk, British Army troops always formed the majority of the garrison. The Royal Artillery had about sixty-five field guns and about the same number of both anti-tank and anti-aircraft guns and it formed the bulwark of the defence. There were about fifty tanks, including over a dozen Matilda heavy infantry tanks armed with two-pounder guns and the thirty Marmon Harringtons of the KDGs.

During April repeated German attacks were just, but only

just, repelled. By May, with the advent of the hot weather and the terrible sand storms, the situation became less hectic and it was decided to reduce the numbers in the garrison. The KDGs were ordered to split up and about 400 men were left in Tobruk under Major Lindsay of 'C' Squadron, while the rest of the Regiment were evacuated by sea to Egypt. The KDGs in Tobruk were divided into three unequal groups; the smallest, serving with light tanks, were attached to 1st Royal Tank Regiment; they left in September when the Australians were relieved by the British 70th Division. Another group was employed as infantry and held a part of the 27-mile-long perimeter. The third group, 'C' Squadron, comprising about ten officers and 210 other ranks, remained, with few changes, throughout the siege. Its main task was to provide two daily patrols of two armoured cars each. One of the patrols was responsible for keeping watch on the western boundary where enemy activity was always greatest. The other was stationed in the flat ground on the southern part of the perimeter where the officer shared an observation post with a Gunner officer in one of the towers built by the Italians. The Squadron also provided a continuous rota of eighteen men who manned six isolated enemy tanks that had been dug in as strong points; they lived in deep slit trenches, a form of shelter that was essential for survival for the heavily bombed and shelled garrison. Nevertheless casualties in Tobruk were remarkably light largely because the soft sand deadened the explosive force of the bombs and shells. To counteract the monotony, Lindsay organized intensive training programmes, which took place in the coolness of the early morning, while in the afternoon many went swimming at the nearby beaches. Space prevents more detail about life in Tobruk, but a full and fascinating account can be found in the *History of the King's Dragoon Guards 1938-45.*

Just before a major German offensive was due to be launched, the Tobruk garrison was ordered, on 21 November, to make their long-awaited break-out to link up with the Eighth Army's offensive, code-named 'Crusader'.

With fifteen armoured cars, Lindsay led the main column to the minefields where the crews, aided by one Sapper per car, lifted the mines to make a gap for the armour to advance. Thirteen cars were either blown up or hit by shell-fire and many of the crews were wounded, but miraculously only one man was killed. Later the losses were much heavier. A week's fierce fighting ensued until the garrison linked up with the New Zealand Division who in turn were temporarily cut off from the rest of the Eighth Army. By 1 December Rommel had begun to retreat and the siege was ended. After eleven and a half months continuous service in the desert and Tobruk, C Squadron drove their remaining battered Marmon Harringtons into Cairo on 20 January, 1942. They were one of the very few units which had spent the whole period in Tobruk and although they had fulfilled their exacting role with great distinction, Lindsay was given no award.

The rest of the Regiment had been equipped with forty-five armoured cars and fought in the 'Crusader' offensive. By late January after Rommel's counter-offensive, they were near Benghazi and thus almost back to where they had been a year earlier. Orders were then received to return to the Nile Delta and the Regiment was at last reunited.

The Bays had spent nearly eighteen months training in England before sailing to the Middle East as part of 1st Armoured Division. Including a welcome stop at Cape Town, the voyage lasted nine weeks and they arrived at Suez on 25 November, 1941. They were now fully equipped with forty-eight tanks, thirty-one being Crusaders, a cruiser tank armed with the inadequate two-pounder gun. The rest were the reliable American light tanks, known as General Stuarts or Honeys, armed with the slightly smaller 37 mm gun. Both these machines were seriously outgunned by the German Marks III and IV tanks which had also much better armoured protection.

The Bays' first experiences in the desert mirrored those of the KDGs. After a short spell for acclimatization and training, 2nd Armoured Brigade was sent 600 miles forward from the railhead at Mersa Matruh to take up the most advanced positions near Antelat, close to the Tripolitanian border. The journey included about 250 miles across track-less boulder-strewn wastes. The sort of problems that the Regiment had to content with on this march were that, on an average, six of the mechanically unreliable Crusaders broke down each day; the Honeys needed refuelling every forty miles, whereas the more comfortable Crusaders had a range of 100 miles; some tanks had no wireless sets and the rubber-tracked Honeys and soft-skinned vehicles had their tracks and tyres continually cut up. Hardly had the Regiment gathered in nearly all its breakdowns when, on 21 January, 1942, Rommel counter-attacked. During the next fortnight the brigade had to withdraw continually and fight a series of dispersed actions, normally on a regimental basis. Through a combination of enemy action, breakdowns and lack of petrol, the brigade had, at one stage, less tanks than a weak regiment and had to form a composite regiment. Nevertheless, by 5 February the Bays were fit for action again, having been supplied with a mixed bag of old patched-up Crusaders and Honeys. By mid-February they were deployed near Knightsbridge, a famous crossroads on what became known as the Gazala Line. They were gradu-ally re-equipped with Crusaders, but also received twelve of the better-armoured and reliable American General Grant tanks whose 75 mm gun could fire both solid shot and high explosive shells and was thus capable of matching the Ger-man armour.

An officer described life in the Desert at that time:

We deployed in 'open leaguer' by Squadrons, each tank being some 200 yards from the next, with longer dis-tances between troops. All the tanks were in sight of the Regimental Headquarters. The day started with a fusillade from the A.A. machine guns mounted on the turrets – partly to ensure that we were all awake, but also to prove that the guns were in working order. At night, when near the enemy, we invariably formed close leaguer. This took the form of a small letter 'n', with one squadron placed across the top in line, the other two forming the legs of the 'n', in single file. If our motorized infantry were present they set up standing patrols at appropriate points and also covered the rear. The replenishment parties came inside the leaguer and remained there throughout the night, whilst the fitters worked on the tanks. No smoking was allowed and we kept as quiet as possible. Latecomers were brought in under the 'toffee apple procedure'. Having identified itself by radio, the lost vehicle would be told to send 'toffee apple' in so many seconds. This meant that it would have to fire the correct Verey Light, saying for instance 'Green – now'. It would then be given over the radio the correct bearing on which to come, i.e. the back-bearing from the leaguer to the vehicle or tank.

At the beginning of May, 1942, the KDGs also moved up to the Gazala Line, having been re-equipped with the latest version of Marmon Harringtons. At the end of this month, Rommel attacked. In the opening stages of the Gazala battle, 2nd Armoured Brigade, of which the Bays formed a part, scored one of the few British successes when they almost wiped out a German Lorried Infantry Regiment. Rommel, however, soon gained the upper hand and within a month had not only decisively defeated the Eighth Army, but had also captured Tobruk. During this prolonged and confusing battle, which swirled across a large area of desert, the British lost cohesion and both Regiments were at times split up and attached to *ad hoc* groups in various sectors of the battlefield.

The Bays were ordered to withdraw to Egypt and the same officer recalled, 'I and my crew climbed on the back of one of the three "runners". I remembered nothing more until first light by which time we were well clear of the Tobruk perimeter; we were lucky not to have fallen off, but sleep seemed then the most important thing'. The Bays had been in action continuously for seventeen days, said to have been a record for an armoured regiment.

In the battle of Alam Halfa (August-September, 1942), the KDGs and Bays each had one squadron involved. The KDG cars were directly linked with Eighth Army Headquarters, giving it continuous and up-to-the-minute information on German tank numbers and movements. By skilfully withdrawing, these troops helped lure the enemy on to the main defensive positions where they were decisively repulsed.

From the beginning of July to the end of October, the Bays were primarily engaged in regrouping, having received 169 men as reinforcements. The Regiment was re-equipped with Crusader tanks, up-gunned with six-pounders, and with the new American Shermans which were a considerable improvement because they had a turret with an all-round traverse, as opposed to the Grants' side-mounted sponson with its very restricted traverse.

The Bays were given a major role in the Battle of Alamein. Montgomery planned that the two British armoured divisions would break through the German lines. These were several miles deep and protected by extensive minefields through which the infantry had first to drive a wedge. 2nd Armoured Brigade were to be the spearhead of 1st British Armoured Division. On 24 October they passed through the British minefields in a cloud of dust and, still in the dark, successfully crossed the first German minefield, but trouble began at dawn as they were going through the lanes of the second enemy minefield whose clearance had been delayed. Another minefield, protected by anti-tank guns, tanks and artillery, lay in front. The Bays soon found themselves unable to move much, let alone advance. A continuous battle raged all day with a major German counter-attack coming in that evening. When night fell, the Bays had lost seventeen out of their twenty-nine Shermans and this was despite gallant work by the fitters who had repaired under fire many of the tanks damaged on the mines. The next day the pattern was repeated, there being an attempt to advance in the morning, which was followed by a German counterattack. A battle of attrition had developed. On 26 October the exchange of fire continued and even after the arrival of eleven replacement tanks that evening, the Regiment was reduced to eleven Shermans and sixteen Crusaders. On 28 October the armour was pulled out, after six days and nights of action.

Montgomery decided to pause before resuming the attack and chose an area slightly to the north of the previous one because more progress had been made there. The Bays were brought up to full strength with twenty-nine Shermans and nineteen Crusaders but ten of these mounted the old two-pounder gun. On 2 November the offensive began again and, after two days intensive fighting, it became apparent that Rommel's forces were defeated. With only twelve Shermans and twelve Crusaders, the Bays drove through the enemy lines and saw the terrible damage that had been inflicted on the Germans and Italians whose tanks, guns and vehicles littered the desert. Although some strong points held up the advance, by 6 November the Regiment had covered forty miles. The main problem had now changed to one of supply. The drivers of the soft-skinned petrol and ammunition lorries who had worked so valiantly refuelling and rearming the tanks during the battle, now found the huge and often unmarked minefields most confusing. These, together with the occasional enemy attacks, seriously delayed the arrival of the resupply lorries, several of which were lost en route. Thus tanks were immobilized for a crucial period through petrol shortages. (Shermans in the soft going at night were using three gallons of petrol per mile.) A small joint column was, however, formed, which included their six remaining Crusaders, to try to cut the coastal road twenty-five miles away, but it ran out of petrol just before reaching its objective and had to gaze on the tantalizing sight of masses of enemy transport escaping. That night a downpour had bogged all the wheeled transport in the desert sands, but it did not affect Rommel's forces who were using the only road in the area. When plentiful supplies of petrol arrived on 8 November, the enemy had got beyond pursuing distance for the armour. As only few tanks were now required, the Bays were taken out of the line until March, 1943. The Alamein Battle had cost the Regiment thirty-two killed.

In the middle of November, 1942, the KDGs left their dusty camp outside Cairo and at the end of the month had again reached the Tripolitanian border. They were now equipped with forty Marmon Harringtons, variously armed, and eighteen of the excellent new Daimlers which mounted two-pounder guns. For the next three months the KDGs spearheaded the Eighth Army's advance towards Tunisia, being one of the two armoured car regiments in 4th Light Armoured Brigade, which also consisted of a tank and an artillery regiment and an infantry battalion.

Early in December some KDG officers who knew the region well were able to convince senior Army commanders that a wide arc-like sweep south of the border to Mereda, (forty miles inside Tripolitania and on the only coastal road) might cut off the retreating Axis forces which were still holding up the advance at El Agheila. A small party in three jeeps led by Lieutenant Richardson, surveyed, mainly at night, that part of the route that lay behind the enemy lines. He returned undetected to confirm that this hook was feasible. Some days later the KDGs led the Brigade, together with most of the New Zealand Division, across the desert on a two-day march. They got astride the road before dawn. All seemed set for a major success but, as had happened often previously, the British had not committed strong enough forces. The experienced and determined German leaders found a weak spot and over a thousand vehicles and tanks escaped through a gap south of the road. A great opportunity to finish off the main enemy formations had been lost.

After a brief rest, the Regiment was sent on to Buerat where Rommel made his next stand. Arriving there on Christmas Day, the KDGs kept the enemy lines under observation for three weeks. Aerial battles had become more frequent and patrols picked up many pilots who had been shot down in the desert and lost. The next move was to Tripoli, 200 miles to the west. Its speedy capture was essential as all stores were having to come in via Benghazi, 400 miles east of Buerat. This time 4th Light Armoured Brigade was ordered to make a long outflanking march which entailed crossing a virtually trackless belt of hills seamed with ravines. About twenty miles in width, this obstacle stretched inland from the coast about twenty-five miles east of Tripoli, and protected it and its airfields from any large-scale flank attack through the desert. The KDGs spent two days probing up steep gullies before finding a route suitable for the Brigade and the New Zealanders. From the summit of these hills the most forward patrols had to watch helplessly as aircraft took off from the huge Castel Benito airfield to attack the British forces. Once through this obstacle, the KDGs raced on forty miles south-west of Tripoli, taking the enemy on an airfield completely by surprise. With this force in their rear, the Axis troops decided to pull back as quickly as they could, and Tripoli fell.

At the end of January, 1943, the Regiment had a brief rest, spent in maintaining and repairing the armoured cars which had had a terrible hammering, only thirty out of fifty-eight being serviceable. Tyre wear was particularly heavy, machine-gunning by aircraft adding to the toll of burst tyres. With the dumps getting further back, petrol supply became a major problem; ten lorries carrying 500 gallons each were needed to sustain the advance. The Regiment's lorries were breaking down frequently and it was difficult to find or borrow replacements. In this more fertile countryside, however, both food and water were relatively plentiful. (The desert ration had been half a gallon per man per day.) Wine was obtainable and there were even houses to sleep in.

Soon on the move again, the KDGs crossed the border into Tunisia on 6 February as the leading troops of the Eighth Army. Strengthened by some of the new and heavier A.E.C. armoured cars, the Regiment now had their own battle group with a battery of eight 25-pounder guns, two troops of 6-pounder anti-tank and some Bofors anti-aircraft guns. This enabled them to combat a much wider range of enemy vehicles and weapons than had previously been the case. Advancing on a broad front inland, mainly through desert, they approached the rugged Matmata Hills that formed the southern part of the Mareth Line. Based on these strong fortifications, the enemy oppostion, both on the ground and in the air, made further progress by the light British forces impossible.

After a pause of about five weeks, the Battle of the Mareth Line began and it proved the hardest struggle that the Eighth Army had had since Alamein. More armoured formations were ordered forward, including the Bays whose journey of over 1000 miles, from near Tobruk, took them a fortnight. A frontal attack having made little headway, Montgomery decided on a left hook via El Hamma, where the New Zealanders, with KDGs in front, had earlier nearly broken through. Both Regiments fought in this renewed offensive against the Mareth Line. After the long and arduous approach march, one of the Bays' officers was greatly encouraged to see the familiar figure of General Horrocks, commanding X Corps, in his tank on the most forward position. During this battle the Bays had the eerie experience of a moonlight attack through the enemy lines, in which they moved at about 2½ mph, shooting up transport and creating chaos. On 29 March the Germans and Italians started to pull back, and early in April Eighth and First Armies finally linked up. Nevertheless during the next six weeks the Axis forces continued to stage tough rearguard actions, in which the Bays had thirteen soldiers killed.

Many Eighth Army units were now being withdrawn to prepare for the Sicilian landings, but not those in which the Regiments were serving and they were therefore transferred to First Army, which occasioned some rude comments! On 13 May nearly 250,000 German and Italian troops laid down their arms, a total only exceeded by that at Stalingrad – many Germans referred to the event as Tunisgrad. Both Regiments now found themselves at the base of the Cape

Bon Peninsula. The masses of prisoners and the huge quantities of equipment gave visual proof of this great victory which was the climax of a long and dramatic campaign. It had taken them nearly 2,500 miles, backwards as well as forwards, across some of the most inhospitable deserts in the world. They now spent a well-earned and peaceful summer near Tripoli.

9

Italy

ON the whole the Italian Campaign was a frustrating one and this was particularly the case for armoured formations. Unlike North Africa, the much more rugged Italian terrain was usually unsuitable for their employment and more armoured units were available than could be used in Italy or elsewhere.

On 24 September the KDGs landed at Salerno and from there one of the Squadrons 'jumped the gun' and liberated Naples. Brigadier Llewellen Palmer later described how this happened:

> On 30 September 'A' Squadron had had a frustrating day trying to advance from Pompeii towards Naples; with communications being so difficult, we had switched off the rear link set (to Regimental HQ). We had seen our chance. Black diesel smoke was rising from the docks and there was unnecessary hostility by the German forward troops. (The Germans never could bring themselves to leave unexpended ammunition behind). 'B' Squadron had been ordered to halt, but 'A' Squadron was still out of contact. We always obeyed direct orders – but if you did not receive any orders you did your best. Three days before, we had spent a very convivial evening with four US reporters and one of them had bet me that the US Army would be in Naples before us and I took the bet. The first of October broke clear. Vesuvius smoked away on the right and black smoke still rose from Naples. When light enough to see clearly, the Squadron started

through the outskirts of the city to find it deserted. Odd figures occasionally darted from one back alley to another and there was an eerie silence. The Squadron was installed in the main square when suddenly the Neapolitans realized that these were British armoured cars. The place erupted into a seething mass of yelling, screaming, gesticulating humanity. Pretty girls hugged and kissed us, and wine, fruit, flowers and vegetables suddenly appeared. Patrols were sent out and soon reported that the city' was free except for the northern approaches. General Mark Clark warmly congratulated the Squadron and I handed over to the Americans on 3 October and was later paid by the American reporters.

It is, however, doubtful whether Mark Clark ever forgave or forgot this well publicised British success. When in May, 1944, General Alexander ordered the American forces in the Anzio bridgehead to break out north to cut off the retreating Germans, Mark Clark disobeyed the order. This time it seemed that he was determined that there should be no mistake and that the Americans should liberate Rome with himself at their head (shades of the novel *Catch 22*). Thus tens of thousands of German troops were allowed to escape to continue the war on the Gothic Line during the dreary winter of 1944-5.

From October, 1943 until March, 1944, some, or all, of the squadrons of the KDGs had prolonged periods of duty as infantry on the Garigliano and Volturno river fronts. In dreadful weather these operations entailed much patrolling, with the inevitable losses from mines. Some horses were acquired and were formed into an unofficial horsed troop which remained a treasured part of the Regiment until after the end of the war. Late in April the Regiment was moved from the western to the eastern flank, coming again under the Eighth Army on the Sangro River. Here they stayed in a predominantly infantry role until, in mid-June, they reverted to being an armoured-car regiment in pursuit of the Germans who were retreating from the Gustav to the

Gothic Line. In this advance, often delayed by blown bridges, the first major town taken was Perugia. In the middle of August they spent a few days in Florence, (declared an open city) with part in British and part in German hands and no fighting permitted by day.

For the next two months they were engaged in sporadic and usually unpleasant actions as the Allies slowly penetrated into the central sector of the Gothic Line. In November they began their last tour of duty in Italy on the Adriatic near Ravenna, having by then changed command thirty times during their fifteen months in Italy, and been in action directly against the enemy longer than any other regiment in the Army since the outbreak of war. Late in December the KDGs sailed to Greece to help suppress the Communist uprising and there they remained until April when they returned to the Middle East.

In the Second World War the Kings Dragoon Guards had lost eighteen officers and 107 other ranks killed.

In May, 1944, the Bays arrived in Italy, still with 2nd Armoured Brigade, and were equipped with Sherman tanks. Early in September they went into action at the eastern end of the Gothic Line in the foothills of the Apennines, between San Marino and Rimini. The countryside here is very broken with steep valleys and irrigation ditches; the few small roads and tracks soon became extremely congested, making resupply and movement in general a laborious and uncertain task. It was therefore most difficult terrain for tanks to support infantry and vice versa. After three days forced march, the Regiment was, on 20 September, ordered to cross the Coriano Ridge before the infantry had cleared the area, and move into the narrow valley that leads down from it. With excellent observation posts on the hills above, the Germans had set up a screen of anti-tank guns (including 88mm) and were waiting for 'B' and 'C' Squadrons. They lost all but three of their tanks in a matter of minutes. This fighting cost the Bays ninety-eight casualties; on Coriano Ridge alone five officers and seventeen other ranks were killed, a total only exceeded in the much more prolonged

Gazala and Alamein battles.

For the next three months the Regiment was engaged in a series of harsh but unspectacular limited offensives. These were aimed at eventually taking Bologna from the east and thus breaking the main Gothic Line positions. These operations had to follow the rugged base of the Apennines. Being always against the grain of the country, the advance was slow as it involved a series of contested river crossings. In mid-October the Bays had the doubtful pleasure of crossing the Rubicon which, unlike most of the other rivers, was not then in full flood. Two months later the Regiment had crossed the Lamone River and although Faenza had been taken, Bologna was still a long way off. The exhaustion of the troops, the shortage of ammunition, the terrible weather and the dogged German resistance decided General Alexander to stop this offensive. For the next three months most of the Bays were dismounted and had monotonous spells of infantry duty on the Gothic Line.

In March the Bays were pulled out of the line to prepare for the Eighth Army's final offensive which began on 9 April. The key to the defences of the Po Valley was the broad, shallow Lake Comacchio which almost merges into the Adriatic. Along its western side the Germans had flooded extensive tracts of land and the only dry strip was known as the Argenta Gap. Before reaching it, the Bays had some hard fighting, including crossing the large Santerno river. They worked closely with leading infantry, giving valuable support, destroying strong points in farmhouses and shelling enemy positions in the vineyards that cover much of this plain. The Regiment also played an important part in the three-day battle to force the Argenta Gap itself. Even after this there was some bitter, sporadic fighting as the British forces pushed north to Ferrara where the Bays found themselves on 2 May when the Germans surrendered.

In the Second World War, the Bays lost twenty officers and 143 other ranks killed whilst serving with the Regiment, of whom nearly one-third, nine officers and 44 other ranks, were lost in less than eight months of the Italian Campaign.

10

The Post War Years

After the war the KDGs moved to Libya, returning to England early in 1948 for one year before going to Omagh in Ulster for a fairly peaceful three years. Their next move was to Germany in BAOR, as a part of British NATO forces, where they spent four years, first at Hamburg and then at Neuminster. After a brief spell in England in 1956 they were stationed in Malaya for the next two and half years. At the time of their arrival in June, 1956, there were two Armoured Car Regiments, equipped with Ferrets and Daimlers, operating in the Malay Peninsula. By this time, General Templer's policy of curfews and fortified villages was beginning to show results. Chin Peng's Communist Terrorist Organization, numbering about 7000 active jungle fighters, was running short of supplies and finding it increasingly difficult to penetrate and subvert rural communities. Nevertheless the Communists were well armed with an assortment of British and Japanese weapons and were still capable of mounting ambushes and specific operations in considerable strength at times and places of their choosing.

The KDGs arrived in Malaya at War Establishment, some 750 all ranks, with a strong Assault Troop element. For the first year, they were deployed in support of the Gurkha Division in the southern half of the country, with RHQ, 'B' and HQ Squadrons at Seremban: 'A' Squadron at Johore Bahru and 'C' Squadron at Kuala Lumpur. In addition, the so-called 'Training Squadron' on Singapore Island was available and frequently used in an internal security role in the City, notably so during the serious riots on 25 October,

1956, when the Squadron was engaged for five days patrolling and enforcing the curfew. The Training Squadron was composed of one Sabre Troop, plus an Assault Section, from each Sabre Squadron, rotating with its own vehicles every two months or so. This practice was continued throughout the tour and proved an extremely effective method of extracting troops from the jungle and maintaining a high standard of discipline, re-training and trade testing.

The Regiment was employed principally on convoy and VIP escort duties and the mileages covered were phenomenal. However the Assault Troops were also engaged on jungle patrols and ambushes. A patrol led by a corporal of 'B' Squadron was the first to draw blood, killing a notorious terrorist and wounding another who stumbled into their ambush position at night. On another occasion, the story goes that a Squadron Leader recommended a young NCO who had done particularly well on operations for a gallantry award. To his dismay, his recommendation was returned by RHQ with a cryptic note from the Colonel which ran, 'Who d'you think we are? The . . . . Brigade?'

Early in 1957, the security situation had so improved that it was decided to reduce the Malayan Garrison to only one Armoured Car Regiment and to form a Federation Recce Regiment based in Kuala Lumpur to which several young KDG officers and NCOs were seconded. Thus, when the 15th/19th Hussars completed their tour in May, 1957, the KDGs were redeployed to cover the whole country. 'A' Squadron and the Training Squadron remained at Johore Bahru and Singapore respectively. 'B' Squadron moved south from Seremban to Kluang: 'C' Squadron moved from Kuala Lumpur to Ipoh in the north of the country, where they were joined by RHQ and HQ Squadron. Yet another detachment was formed, The Gurkha Div HQ Escort Troop, consisting of ten Ferret Scout Cars which remained in Seremban. Thus the Regiment was split up over an area the size of England in Troop and Squadron detachments – an RSM's nightmare and a Troop Leader's paradise. Soldiers have seldom been so happy and full of purpose, before or

15 A Recce Flight helicopter and a Saladin armoured car on exercises in Germany, 1962.

16 Saracen and Saladin in Aden, 1967.

17 The Regiment on Parade, Germany, 1973.

18 & 19 In Northern Ireland, 1976.

since. Admittedly terrorists were becoming a trifle overshot, but as the threat to communications receded, troops moved deeper and deeper into the 'ulu'.

'C' Squadron now supported the Commonwealth Brigade (1 Loyals, 3 Royal Australian Regiment, 1 NZ Regiment, and 2/6 Gurkhas). They also assumed the happy task of providing a detachment in the Cameron Highlands, a hill station and leave centre at an altitude of 6000 ft. While the rest of the Squadron combed the jungles of Perak, any man in need of a change of air was sent up to the Cameron's detachment for a breather. All vehicles, civil or military, proceeding to the Highlands were required to travel in convoy and every day, seven days a week, the detachment escorted a morning and evening column up and down the tortuous 40-mile road from the Highlands to the Plains. Latterly, this road became an accepted place for demoralized terrorists to surrender and the KDGs took over 70 on this road alone before their departure. The officer in charge there became a local legend in the Camerons and led some intrepid patrols to remote aboriginal settlements in search of the enemy.

On 31 August, 1957, the Federation of Malaya was granted independence. 'B' Squadron represented the Regiment at the *Merdeka* (Freedom) Parade in Kuala Lumpur, at which HRH The Duke of Gloucester took the salute.

In October, 1958, they handed over to the 13th/18th Hussars (QMO) and, among a spate of valedictory telegrams, perhaps one from the Commander-in-Chief Far East, Lieut-General Sir Richard Hull, best summed up the Regiments achievements:

> During the last two and a half years your tasks were to keep the roads open throughout the Federation for the safe passage of Security Forces and Civilians and to give aid to the Civil Power in Singapore in time of need. Both these tasks you have accomplished magnificently. Your vehicles travelled some three million miles on patrol and escort duties and as the Communist Terrorist threat outside the jungle decreased you went on

foot into the jungle in search of the enemy. You have earned the unbounded respect and admiration of all units of the Security Forces and of the Federation and Singapore Police Forces.

You go home now to amalgamate and I know that you will take with you to 1st The Queen's Dragoon Guards the same efficiency, high spirit and enthusiasm which you have shown here. Good luck to you all.

'A' and 'B' Squadrons embarked at Singapore on 18 October and the remainder of the Regiment at Penang on 20 October to set sail for home by way of Colombo, Aden and Suez. They docked at Southampton on 11 November to find the Band and Trumpeters of the The Queen's Bays on the quayside to greet them, and the following evening the Regiment was comfortably bedded down in The Bays Barracks at Perham Down. The Bays disappeared on leave the following day while the KDGs prepared for their Final Parade.

At the end of the Italian campaign the Bays remained in Northern Italy, staying there until the early summer of 1946. They then moved to the Canal Zone. As a result of the British Government's decision to withdraw from Palestine, the Regiment returned home to England in time for Christmas of that year after an absence of six and a half years. They were stationed at the Dale, Chester for nearly two years during which time they were temporarily reduced to cadre strength.

In 1949 the Regiment, under command of Colonel K. E. Savill, was honoured to receive a visit by the Colonel-in-Chief. It was the first time that many of the post-war Regiment had been privileged to meet Queen Elizabeth so that it was a particularly memorable occasion. By the end of the year the Bays, now back to full strength, had moved to Fallingbostel where they served under command of the 7th Armoured Division as an Armoured Regiment. After a five year spell in Germany the Regiment was warned for service in Korea, but the decision to reduce the size of the Commonwealth Division meant that this was changed, much to the disappointment of many. Plans were then laid for the Bays to spend two years at Tidworth and they left Germany in September, 1954. However, immediately on arrival in England, Colonel Manger, the Commanding Officer, was told that the Regiment was to leave in three months time for the Middle East. Little could be done in this short time apart from proudly welcoming a visit by the Colonel-in-Chief.

In December, 1954, the Regiment left Southampton by troopship for Aqaba in Jordan. On arrival the Regiment was widely deployed. One squadron was located at Ma'an with another in Iraq where they were engaged in training the Iraq Army in the use of the Centurion tank. The remainder of the Regiment was at Aqaba. A further training commitment in the shape of assistance to the Jordan Army followed shortly.

In September, 1955, 'C' Squadron moved to Sabratha in Libya, some 50 miles to the west of Tripoli. The Regiment was thus widely dispersed and with two regiments' worth of tanks to maintain – no easy task.

In February, 1956, the Regiment moved to Sabratha to join up with 'C' Squadron. Sabratha proved to be a good, if rugged, station. Situated on the shores of the Mediterranean it was adequately served with training areas and firing ranges, while there were ideal opportunities for all kinds of sport and recreation.

However, the Regiment's peacetime preoccupations were dramatically shattered in the summer of 1956 by the events leading up to the 'Suez Affair'. One evening while the Regiment was on training in the Desert well to the south of Sabratha, a liaison officer arrived from Divisional Headquarters with a message for Colonel Armitage, the Commanding Officer, that the Regiment was to return to barracks and prepare for war.

Preparations were immediately set in hand to bring all equipment up to battle efficiency. Reinforcements in the shape of reservists started to arrive almost immediately.

Naturally the reservists required refresher training so everybody was very much on the go. Frequent regimental exercises became a matter of routine.

The original plan had been for the 10th Armoured Division, of which the Regiment formed part, to move overland to Alexandria. This would have been a long haul by any reckoning. With tanks of uncertain age and without the assistance of tank transporters this was an ambitious, if not over-optimistic, plan. There was one additional snag. Nobody had consulted the Foreign Office. When this was finally done the Foreign Office pointed out that the Treaty by which British Troops were in Libya precluded such a manoeuvre. Plans were subsequently changed and the 10th Armoured Division were no longer to lead the assault. If required it would move by sea and form part of the follow-up force.

Meanwhile local unrest in Libya was increasing. The Regiment had many families living in private accommodation in Zavia, a Libyan village 15 miles from Sabratha. For their safety these families had to be brought into the camp at Sabratha; this involved a lot of dislocation and doubling up which was accepted with good humour and surprisingly little complaint. The British families in Tripoli had already been flown home so there was a firm incentive to make the best of a difficult situation.

All locally employed civilian labour had to be denied access to the camp for security reasons. This meant that regimental wives and children were helping out in the NAAFI, the camp cinema and elsewhere.

Regimental patrols were in constant operation, not only for protection but to oversee certain farms which were suspected of concealing arms and ammunition. Even the polo ponies were given an operational role by their use for certain mounted patrols.

The outcome of the 'Suez Affair' is well known and certainly forms no part of this history, nor is it appropriate to comment upon the political wisdom of Government action.

In the event the Regiment was never used, apart from a

squadron involvement in internal security tasks in Tripoli. Slowly life returned to normal although the atmosphere throughout Libya was never quite the same. The reservists returned to civilian life a few weeks later. They had responded most willingly and cheerfully to all demands made upon them and had quickly relearnt the skills of tank soldiers.

Upon the 1957 reorganization of the Army the Regiment learnt that it was to amalgamate on 1 January, 1959, with the KDGs. If amalgamation it had to be, then no happier choice could have been made. It also had been decided that the British Army would give up the military station of Sabratha. In the autumn of 1957 the Bays, therefore, returned to England.

The Queen's Bays spent their last year of individual existence at Perham Down, near Tidworth. During this time they converted to the armoured car role in preparation for amalgamation and were actively involved in assisting at Yeomanry camps at Castlemartin during the summer months.

On 1 November, 1958 the Regiment paraded for the last time, and was reviewed by Her Majesty Queen Elizabeth The Queen Mother, the Colonel-in-Chief. The place chosen for this great occasion was the Tattoo Ground at Tidworth, not so very far from Salisbury where Historical Records show King James II reviewed the Regiment on 21 November, 1688. It was an incomparable setting, and one which brought back memories to old Bays of taking part in the Tattoo in years gone by, and more particularly of the parade on 25 July, 1939, when the Regiment, mounted in Mark VI Light Tanks, was presented with a new Standard by the Colonel-in-Chief.

The Regiment paraded in Armoured Vehicles with the Standard Party mounted on horses. During the Colonel-in-Chief's speech, Her Majesty said, 'One of my earliest duties after assuming my appointment was to present the Standard which is carried today on parade for the last time. That occasion, which I remember so vividly, took place on this same ground just five weeks before the outbreak of war. Your achievements since that day have added another splendid chapter to the history of the Regiment'.

After the parade Her Majesty inspected the very fine gathering of Old Comrades under command of Lieutenant-General Sir Wentworth Harman. It was a sad and moving occasion for all past and present members of the Regiment.

On 2 March, 1959, Her Majesty presented the new Standard emblazoned with these battle honours:

BLENHEIM, RAMILLIES, OUDENARDE, MALPLAQUET,
DETTINGEN, WARBURG, BEAUMONT, WILLEMS, WATERLOO,
SEVASTOPOL, LUCKNOW, TAKU FORTS, PEKIN 1860,
SOUTH AFRICA 1879, SOUTH AFRICA 1901-1902

MONS, LE CATEAU, MARNE 1914, MESSINES 1914, YPRES 1914-5,
SOMME 1916-18, MORVAL, SCARPE 1917, CAMBRAI 1917-18, AMIENS,
PURSUIT TO MONS, FRANCE AND FLANDERS 1914-1918.
AFGHANISTAN 1919.

SOMME 1940, BEDA FOMM, DEFENCE OF TOBRUK, GAZALA,
DEFENCE OF ALAMEIN LINE, EL ALAMEIN,
ADVANCE ON TRIPOLI, TEBAGA GAP, EL HAMMA, TUNIS,
NORTH AFRICA 1941-3, MONTE CAMINO, GOTHIC LINE,
CORLANO, LAMONE CROSSING, RIMINI LINE, AGENTA GAP,
ITALY 1943-45.

II

1st The Queen's Dragoon Guards

T he two Regiments were amalgamated on 1 January, 1959, as a Reconnaissance Regiment equipped with Alvis Saladin armoured cars. Later that year they moved for over five years to Wolfenbuttel and this was followed by nearly two years in Omagh, from where first 'B' Squadron and then 'C' Squadron were sent for six months to Borneo. They were accompanied by their Air Section which was at first equipped with two Auster IX light aircraft and then later by two Sioux helicopters. On the ground, the Squadrons had Saladins, Saracen armoured carriers and Ferret scout cars. In 1965 a sporadic guerrilla war, known as Confrontation, was still being waged in northern Borneo by Indonesia, then under the rule of President Sukarno. His object was forcibly to detach Sabah (formerly North Borneo) and Sarawak from the newly independent state of Malaysia which then had no army of its own. The 1000-mile border with Indonesia consisted mostly of rugged mountains covered with forests and intersected by rivers.

The Squadron was based at Engkilili in Sarawak under command of Mid-West Brigade (19 Infantry Brigade), with detachments in nearby battalion forward bases. The primitive basha huts and dugouts themselves became a forward operation base, from which the whole range of armoured car operations took place, except reconnaissance. The Saladin 76 mm gun and its machine gun were used to support infantry patrols on border operations; road supply convoys between Kuching and Simmangang were escorted; Saladins were misused as static fire platforms within infantry forward bases and many rounds were fired at suspected Indonesian columns in the nearby jungle; and the squadron's air troop was used extensively on border surveillance. In their last 2 months, each squadron was allowed to take its place alongside infantry battalions in border activities, and a programme of foot patrols and ambushes of up to five days duration led to much excitement but sadly no contacts. Visits to isolated tribes were also made as part of the 'Hearts and Minds' campaign of psychological warfare. This was aimed at winning the confidence of the local inhabitants, thereby preventing them assisting the enemy and perhaps even obtaining from them warnings of future raids or ambushes. This detached role was considered to be a very worthwhile task, rewarding, exciting and thoroughly different from the BAOR/Omagh life.

From August 1966 to July 1967, the whole Regiment was stationed, probably for the last time in its history, outside Europe. It was based in Aden Colony, with a Squadron at Sharjah on the Persian Gulf. This detached Squadron's task was a relatively peaceful one and gave the opportunity for Squadrons in Aden to be changed over and have a break from the heavy commitments in the Colony. The Regiment inherited a collection of old vehicles, including Saladins, Saracens and Ferrets and everyone did a wonderful job to keep them running.

An officer wrote of this period:

> Saladins gave static day and night fire support from well sighted and sandbagged sangers within the perimeter defences of up-country parts such as Dhala and Habilayn and endless convoys were guarded during their moves to and fro on the Dhala 'road'. As the internal situation deteriorated the Squadrons were transferred from their up-country bases, near the border with the Yemen and from their desert patrols, to help in urban operations. Some notable successes had been achieved in preventing arms, ammunition, mines and grenades from reaching Aden from inland.

The Regiment's helicopters proved invaluable throughout these widely dispersed operations. A strange event was the interception of a small motorboat with three deserters from the French Foreign Legion who were shipwrecked on the Aden Coast after crossing the Red Sea from Djibuti.

By early 1967 urban operations had increased in intensity and strikes and industrial action by the various local unions regularly brought Aden Port to a standstill. Once the British Government had announced its intention of quitting the Colony by the end of 1967, fierce fighting broke out between the various rival terrorist groups in order to gain the upper hand in time for independence.

At first, it was felt that, because of the danger of Molotov Cocktails and hand grenades, armoured cars would be too risky to use in built-up areas and in the shanty towns of Aden and Sheik Othman. However, the Regiment managed to prove that armoured car patrols could be used with success, with considerable relief to the hard-pressed infantry on their feet or in their un-armoured vehicles. Violence grew in intensity with the coming of the hot weather and by the middle of April, 1967, the Regiment had been engaged in at least three large-scale street battles, the Battle of the Sheik Othman Mosque being the most lively. Lessons from the past were re-learned. Ferret Scout Cars were easy prey for anti-tank mines and could not be used singly. Six-wheeled Saracens and Saladins on the other hand survived well, and, despite the loss of a wheel or two, were, in most cases, able to motor on after being blown up on anti-tank mines. Land Rovers were weighed down with heavy sand-bagged floors and large roll bars were fitted. Many hours were spent in routine patrolling, manning check points, cordons and searches and endless checks on vehicles and individuals. However, there was still time for bathing, athletics, sailing, aerial shark shooting, deep sea fishing and polo, plus various

99

sightseeing trips to off-shore islands.

The terrorist campaign crippled the intelligence services and, by the end of June, the Special Branch had virtually ceased to exist and the loyalty of the Aden Police force had been destroyed. But a hand grenade, thrown by a policeman, had caused the only serious casualty sustained by the Regiment. The crisis came quite suddenly, on 20 June, during the routine battalion hand-over when the Company Commander responsible for the old Arab City in the Crater district was driving his relief round in an open Land Rover. The local armed police, who had become disaffected, opened fire at point blank range from the roof of their barracks and during this and other engagements on that day a total of twenty-two British soldiers and officers died and thirty-one were wounded.

'A' Squadron was under command of the Crater Battalion and they were rapidly involved with trying to rescue the wounded, recover the dead and contain the fighting to the local area. The action first started outside Crater when two Troops were deployed to assist an infantry Company secure the armoury where Arab troops had mutinied and were firing on passing British Army vehicles, causing numerous casualties.

The armed police, who had heard of the fighting, reacted by firing on the nearest British soldiers that came their way. Another infantry Company began to deploy their OPs in Crater and a QDG Troop assisted by patrolling in two half-troops. A Regimental helicopter which was deploying soldiers to an Observation Post on Temple Cliffs on the Crater lip overlooking the police barracks now came under fire, was hit and crash-landed on the steep rocky slope. One passenger was uninjured and he dragged the others clear before the aircraft burned out.

Meanwhile the Troop in the Crater was trying unsuccessfully to recover eight bodies from the road-way outside the armed police barracks, but were prevented from doing so by heavy small-arms fire and the receipt of orders forbidding fire to be opened on any Arab troops or police. Another Troop also tried, but was forced to withdraw for the same reasons, but not before suffering two casulaties. It was a sad day for the two battalions in the Crater and many felt resentful, regarding the orders as too restricting and feeling that a show of force, and in particular the prompt use of the Saladin's main armament, would have brought the situation rapidly under control. As it was Crater was evacuated and was not re-occupied by British troops for a further three weeks. After this stalemate the relieving battalion, with 'A' Squadron in front, re-entered Crater; the occupation was quick, bloodless and efficient. It was now time for the Regiment to leave Aden.

A short posting in England was largely spent in converting to an armoured regiment and once again it was off to Germany, this time for nearly three and half years at Detmold. The QDGs next became the Royal Armoured Corps training regiment at Catterick for two years, with 'A' Squadron detached in Berlin. In January, 1973, the Regiment was back in Germany at Hohne and was equipped with Chieftain tanks. In the Army the 'cavalry' routine now seems fairly well defined. Most of the time is spent in Northern Germany as an armoured regiment in BAOR, interspersed with shorter spells of about two years in England either as a demonstration or a training regiment. Less frequently the Regiment will again become a Recce unit, either in BAOR, or in Northern Ireland or in England with Squadrons serving in Cyprus.

Finally the cavalry regiments now take their turn to go to Ulster for four-month unaccompanied tours, largely in an infantry role, and the QDGs have recently been there twice. (Between 1689 and 1975 the two Regiments had spent a combined total of nearly 70 years in Ireland). Operating in Ulster usually entails being split into small detachments, especially in rural areas; indeed it has been termed an NCO's war. This account, by a Lance Corporal well described the conditions in December, 1975;

> Coalisland Security force consisted of a number of RUC policemen, all living in the Station, a Section of eight QDGs under a Lance Corporal, the Permanent Detachment Commander, a Sergeant from the QDGs, and his assistant, a Lance-Corporal. The sections were detached from the Squadron HQ at Dungannon for a two-week tour of duty. We all lived in the 100-year-old building which had all its windows shot out. Heavy armoured shutters were mounted over its windows and doorways. Outside, an anti-mortar wire protected the inmates from any likely threat from that angle.
>
> It would be incorrect to say that a routine was soon set, because one had to avoid routine action like the plague. All patrols, whether on foot or mobile, would never cover the same area the same way. One had to play a game of wits, trying to outfox and surprise the opposition at all times. Jumping over walls, appearing suddenly in the town square, driving round the country lanes with no lights, dropping off foot patrols, returning to the abode. All the time, the Section's comings and goings were observed from the cafe across the way from the Station. Most people enjoyed their stay in the centre of this Republican village, working long hours under varied conditions, where it was felt that the Section had gained the respect of friend and foe in the area. The Section matched the local wall artist who insisted on painting slogans of an unpleasant nature on the largest wall available, the answer was merely to build the wall bigger and put up better slogans. Greetings were given to everybody – not all were returned.

Epilogue

The domestic life of a regiment is hard to chronicle adequately. Like a happy marriage, the internal affairs of a good regiment are unobtrusive but contented. The total number of wives and children is now about equal to that of all the serving soldiers. It is a young community whose needs and welfare demand constant care and attention, with the Regiment moving often and sometimes at short notice. No civilian organization has to be so self-contained, nor does it have to face the constant problems of separation and danger that have become a part of Army life as a result of the disturbances in Ulster. Although usually stationed in a foreign country, these and other complications are accepted philosophically by most familes and tend to help make the Regiment into a close-knit society. Through the Old Comrades Association, this family atmosphere continues even after soldiers and their dependents have retired, thereby linking the younger with the older people who are remembered and, if necessary, assisted, should misfortune befall them. The excellent, small Regimental Museum in Shrewsbury forms another bond, visibly recording the past traditions and glories of the Regiments and is situated in the capital of the Welsh border from where most of the recruits are drawn. Finally, a well-illustrated *Journal* acts as a further connexion between all members past and present of the Regiment.

For over a century sport has rightly occupied an important and continuous part in the life of soldiers. The parent regiments and the QDGs have, and have had, many outstanding team and individual performers and always had a tradition of success, especially in equestrian events.

On the professional plane, the Regiments have been rightly proud of their more famous soldiers. A very creditable number from such a small organisation have risen to a high rank, in particular the Harman family. James Wentworth Harman went to France in 1914 in Command of 'C' Squadron, The Queen's Bays. He was badly wounded in the Retreat from Mons, not returning to France until 1915 when he was appointed to command the 18th Hussars. By the end of the war he was in Command of the 3rd Cavalry Division on the Western Front and Haig said of him that he was 'the best of the three Cavalry Divisional Commanders'. In 1926 he became Inspector-General of Cavalry. He commanded the 1st Division from 1930 to 1933. He was Colonel of The Queen's Bays from 1930 to 1945. He had the distinction of having led in action every unit and formation of cavalry from a Troop to a Division. His son joined the Queen's Bays in March, 1940, and was a Squadron Leader during the Second World War. He commanded the 1st Queen's Dragoon Guards from 1960 to 1962. His subsequent appointments have included Command of the 1st Division in BAOR (some 40 years after his father). Commandant of the Royal Military Academy, Sandhurst and Commander of the 1st British Corps; he is currently Adjutant-General. Since 1975 he has been Colonel of 1st Queen's Dragoon Guards.

An intimate link was recently forged with the Royal Family when Captain Mark Philips (son of Major P.G.W. Philips, who had served with the KDGs in the war, and grandson of Brigadier Tiarks, Colonel of the KDGs from 1953 to 1959 and the first Colonel of the QDG) married Princess Anne in Westminster Abbey on 14 November, 1973. The Regiment provided the Guard of Honour and the bridegroom and best man wore the new dress uniform of the QDGs for the first time.

The Establishments and Costs of the Regiments

Published on 1 January, 1686, the official manual sets out the composition and costs of the regiments, that of the Third Regiment of Horse was as follows.

Field and Staff-Officers	*Per Diem*		
	£	s	d
The Colonel, as Colonel	0	12	0
Lieutenant-Colonel, as Lieut-Col	0	8	0
The Major, who has no troop, for himself, horses and servants	1	0	0
Adjutant	0	5	0
Chaplain	0	6	8
Surgeon, 4/- per day and a horse to carry his chest 2/- per day	0	6	0
A Kettle-Drummer to Colonel's troop	0	3	0
	3	0	8

The Colonel's Troop			
The Colonel, as Captain, 10/- per day, and two horses, each 2/- per day	0	14	0
Lieutenant 6/- per day and two horses each 2/- per day	0	10	0
Cornet 5/- and each horse 2/- per day	0	9	0
Quarter-Master 4/- and one horse at 2/-	0	6	0
Three Corporals each at 3/- per day	0	9	0
Two Trumpeters each at 2/8 per day	0	5	4
Forty Private Soldiers each at 2/6 per day	5	0	0
	7	13	4

Five more troops of the same number and at
  the same rate of pay as the Colonel's
  troop   38   6   8
                                                  _____

Total daily pay   49   0   8
                                                  _____

Total annual pay   £17,897   3   4
                                                  _____

With its nine troops, the total daily pay of the Queen's
Regiment of Horse was £72. 0. 8. and its total annual pay was
£26,292. 3. 4.[1]

**In 1747 the establishment and cost of the KDG was
as follows:**

Staff Officers

	Per Diem		
The Colonel, as Colonel 15/-, allowance for servants 4/6	£ 0	19	6
Lieutenant-Colonel, as Lieutenant-Colonel	0	9	0
Major, as Major	0	5	0
Chaplain	0	6	8
Adjutant	0	5	0
Surgeon	0	6	0

[1] The figures for rates of pay are taken from Richard Cannon 'Historical
Records of the First or Kings Regiment of Dragoon Guards'. p.7.

The First Troop

	Per Diem		
Captain 8/-, 3 horses 3/-, in lieu of servants 4/6	0	15	6
Lieutenant 4/-, 2 horses 2/-, servants 3/-	0	9	0
Cornet 3/-, 2 horses 2/-, servants 3/-	0	8	0
Quarter-Master for himself and horse 4/-, servants 1/6	0	5	6
3 Sergeants, at 2/9 each	0	8	3
3 Corporals at 2/3 each	0	6	9
2 Drummers at 2/3 each	0	4	6
1 Hautbois (Oboe player)	0	2	0
59 Dragoons at 1/9 each for man and horse	5	3	3
Allowance for widows	0	2	0
For clothing lost by deserters	0	2	6
For recruiting expenses	0	2	4
For Agency	0	1	2
Eight more troops of the same Numbers	68	6	0
Total	79	7	11

The annual cost of the Kings Dragoon Guards was £28,979. 9. 7. and of the Queen's Bays, with its three fewer troops, was £19,630. 18. 4.[1]

The three senior staff officers in the regiments each commanded a troop of their own and thus, in addition to emoluments of their own rank, also drew the pay and allowances of a captain. The King had laid down the rates of pay for commissions in the senior cavalry regiments and these were such as to debar all but the very rich from achieving high rank. The following table illustrates the rising expense of a commission during the eighteenth century.

R. Cannon p.60

	1720	**1760**
Colonel and Captain	£9,000	?
Lieutenant-Colonel and Captain	£4,000	£5,350
Major and Captain	£3,300	£4,250
Captain	£2,500	£3,150
Lieutenant	£1,200	£1,365
Cornet	£1,000	£1,102
Adjutant	£ 200	?

1. **PAY AND COSTS of the 1st Queen's Dragoon Guards as at March, 1977**

   Basic Daily Rate of Pay – Other Ranks

WO 1	×	3	@	£11.50	=	34.50
WO 2	×	9	@	£11.00	=	99.00
S Sgt	×	27	@	£10.50	=	283.50
Sgt	×	49	@	£10.20	=	499.80
Cpl	×	105	@	£ 9.50	=	997.50
L Cpl	×	105	@	£ 8.50	=	892.50
Tprs	×	270	@	£ 7.80	=	2,106.00

   Total Daily ROP = £4,912.80

   Total Annual ROP = £1,788,259.20

Basic Daily Rate of Pay – Officers

Lieut-Col	×	1	@	£20.00 =	20.00
Major	×	8	@	£16.00 =	128.00
Capt	×	11	@	£12.50 =	137.50
Lt	×	9	@	£10.20 =	91.80
2 Lt	×	6	@	£ 7.00 =	42.00
		35	Total Daily ROP =		£419.30

Annual ROP	=	£152,625.20

Total Daily Rate of Pay for the Regiment	=	£5,332.10

Annual Rate of Pay for the Regiment	=	£1,940,884.40

These rates are considerably higher when Local Overseas Allowance and other increments are added. The final figure for a Regiment in BAOR would therefore be more than double the £1.9m shown.

Equipment capital costs are approximately £13 m. The main items are 69 tracked and 79 wheeled vehicles.
Equipment running costs per annum amount to about £1.5m.

2. **OTHER FACTS**

Average ages:

Troopers	=	21 yrs
L Cpls	=	22 yrs
Cpls	=	26 yrs
Sgts	=	32 yrs
Majors	=	35 yrs
Capts	=	28 yrs
Lts	=	24 yrs
2 Lts	=	22 yrs

Colonels of the Regiments

1st The King's Dragoon Guards

Lt-Gen.	Sir John Lanier	1685
Gen.	Hon. Henry Lumley	1692
Col.	Richard Ingram, 5th Viscount Irvine	1717
F.M.	Sir Richard Temple, Bt., 1st Viscount Cobham	1721
Lt-Gen.	Henry Herbert, 9th Earl of Pembroke	1733
Gen.	Sir Philip Honywood, KB	1743
Lt.-Gen.	Humphrey Bland	1752
Gen.	John Mostyn	1763
F.M.	Sir George Howard, KB	1779
Gen.	Sir William Augustus Pitt, KB	1796
Gen.	Francis Augustus Eliott, 2nd Lord Heathfield	1810
Gen.	Sir David Dundas, GCB	1813
Gen.	Francis Edward Gwyn	1820
Gen.	William Cartwright	1821
Gen.	Sir Henry Fane, GCB	1827
Gen.	Hon. Sir William Lumley, GCB	1840
Gen.	Charles Murray Cathcart, 2nd Earl Cathcart, GCB (Lord Greenock)	1851
Gen.	Sir Thomas William Brotherton, GCB	1859
Gen.	Sir James Jackson, GCB, KH	1868
Gen.	Henry Aitchison Hankey	1872
Lt-Gen.	Sir James Robert Steadman Sayer, KCB	1886
Maj.-Gen.	William Vesey Brownlow, CB	1908
Lt-Gen.	Sir Charles James Briggs, KCB, KCMG	1926
Brig-Gen.	Alexander Gore Arkwright Hore-Ruthven, 1st Earl of Gowrie, VC, GCMG, CB, DSO	1940

| Brig. | Sidney Howes, DSO, MC | 1945 |
| Brig. | John Gerard Edward Tiarks | 1953 |

Amalgamated with 2nd Dragoon Guards 1.1.1959 to form 1st
The Queen's Dragoon Guards

The Queen's Bays 2nd Dragoon Guards

Col.	Henry Mordaunt, 2nd Earl of Peterborough, KG	1685
Brig-Gen.	Hon. Edward Villiers	1688
Maj-Gen.	Richard Leveson	1694
Gen.	Daniel Harvey	1699
Col.	John Bland	1712
Col.	Thomas Pitt, 1st Earl of Londonderry	1715
F.M.	John Campbell, 2nd Duke of Argyll, KG, KT	1726
Gen.	William Evans	1733
Gen.	John Montagu, 2nd Duke of Montagu, KG, KB	1740
F.M.	Sir John Louis Ligonier, 1st Earl Ligonier, KB	1749
Maj-Gen.	Hon. William Herbert	1753
Lt-Gen.	Lord George Sackville	1757
Gen.	John Waldegrave, 3rd Earl Waldegrave	1759
F.M.	George Townshend, 1st Marquess Townshend	1773
Lt-Gen.	Sir Charles Gregen Craufurd, GCB	1807
Lt-Gen.	William Loftus	1821
Gen.	Sir James Hay, KCH	1831
Gen.	Sir Thomas Gage Montresor, KCH, KC	1837
Gen.	Hon. Henry Frederick Compton Cavendish	1853
Maj-Gen.	Henry Dalrymple White, CB	1873
Gen.	Alexander Low, CB	1874
Maj-Gen.	Thomas Pattle	1881
Gen.	Sir Charles Pyndar Beauchamp Walker, KCB	1881
Gen.	Sir William Henry Seymour, KCB	1894

Lt-Gen.	Sir Hew Dalrymple Fanshawe, KCB,KCMG	1921
Lt-Gen.	Sir Antony Ernest Wentworth Harman,	
	KCB, DSO	1930
Brig.	James Joseph Kingstone, CBE, DSO, MC	1945
Colonel	George William Charles Draffen, DSO	1954

1st The Queen's Dragoon Guards

Brig.	John Gerard Edward Tiarks	1959
Colonel	George William Charles Draffen, DSO	1961
Colonel	Kenneth Edward Savill, DSO, DL	1964
Brig.	Anthony William Allen Llewellen Palmer	
	DSO, MC	1968
Gen.	Sir Jack Wentworth Harman, KCB,	
	OBE, MC	1975

With some corrections, this list is based on N.B. Leslie's 'The Succession of Colonels of the British Army from 1660 to the Present Day', Society for Army Historical Research, Special Publication No.11 1974. The rank given is that held by the Colonel on retirement and not, as in some earlier records, that held on being appointed.

Lt. Colonels (Commanding Officers)

Lt-Col	H C Selby MC	1959-1960
Lt-Col.	J W Harman OBE MC	1960-1962
Lt-Col	P R Body	1962-1964
Lt-Col	T Muir	1964-1967
Lt-Col	G N Powell	1967-1969
Lt-Col	J H Lidsey	1969-1971
Lt-Col	M R Johnston OBE	1971-1973
Lt-Col	R C Middleton OBE	1973-1975
Lt-Col	R W Ward MBE	1975-1977
Lt-Col	J I Pocock MBE	

Brief Bibliography

Beddington, Major-General W. R., *A History of the Queen's Bays 1929-1945*, Winchester, 1954

Cannon, R. *Historical Record of The First, or King's Regiment of Dragoon Guards*, London, 1837

Cannon, R. *Historical Record of the Second, or Queen's Regiment of Dragoon Guards*, London, 1837

McCorquodale, Colonel D, Hutchings, Major, B.L.B. and Woozley, Major, A.D., *History of the King's Dragoon Guards 1938-1945*, Glasgow, 1950

Whyte, F. and Atteridge, A.H., *A History of the Queen's Bays, 1685-1929*, London, 1930